The Great Regeneration

The Great Regeneration

ECOLOGICAL AGRICULTURE, OPEN-SOURCE TECHNOLOGY, AND A RADICAL VISION OF HOPE

DORN COX

with Courtney White

Foreword by **DAVID BOLLIER**

Chelsea Green Publishing
White River Junction, Vermont
London, UK

Excerpt from "Parables" as published in *Everything Speaks in Its Own Way* (2012) by Kae Tempest,
is reprinted with permission of Lewinsohn Literary Agency Ltd.

Acquiring Editor: Ben Watson
Developmental Editor: Ben Trollinger
Project Manager: Rebecca Springer
Copy Editor: Diane Durrett
Proofreader: Angela Boyle
Indexer: Linda Hallinger
Designer: Melissa Jacobson

Printed in the United States of America.
First printing March 2023.
10 9 8 7 6 5 4 3 2 1   23 24 25 26 27

ISBN 9781645020677 (paperback) | ISBN 9781645020684 (ebook)

Library of Congress Cataloging-in-Publication Data is available.

Chelsea Green Publishing
P.O. Box 4529
White River Junction, Vermont USA

Somerset House
London, UK

www.chelseagreen.com

To my family . . .

Contents

Foreword

Of the many epic challenges that climate change is bringing to humankind, one of the most significant is surely the need to reinvent agriculture. Can the world's farmers find a way to shift from large-scale, carbon-intensive industrial farming that is destroying soil and ecosystems to smaller-scale bioregional systems that not only respect nature but regenerate it? Can we invent systems that grow enough nutritious food, distribute it fairly to all, and remake agriculture as a decentralized, place-respecting enterprise?

At this point in the unfolding climate catastrophe, these ambitions are not simply a nice fantasy to ponder. They are existential necessities. If humankind is going to avoid fatal disruptions to the planet's ecosystems and civilization itself, agriculture must find ways to pursue some radical shifts.

In the short term, the top imperative must be new strategies for *adapting* to climate change: new cultivation practices, new crop choices, holistic commitments. Over the longer term, the art of farming must reintegrate itself with local ecosystems and the biosphere. Agriculture must do more than "sustain" an already degraded landscape. It must understand and improve the generativity of life itself.

Dorn Cox offers us a powerful framework for undertaking this task in *The Great Regeneration*, replete with myriad examples of soil restoration and ecological monitoring, farm hacks and open-source observatories, and social and ethical principles for keeping regenerative agriculture on the right track. This book introduces an impressive storehouse of innovations that illuminate many pathways forward.

The Great Regeneration does not provide a blueprint so much as a range of powerful methodological shifts needed to open up new vistas of possibility. With active participation and ingenuity, farmers can begin to take practical steps that draw on recent findings in earth sciences; new applications of open-source software, networking tools, and data systems; bold experiments that blend low-cost observational technologies with attentive human stewardship of landscapes; new organizational forms and cooperative financial models for self-reliance; and patterns of commoning that empower individuals and communities.

Regeneration, as Cox points out, is not simply a set of techniques. It is a mindset and worldview. It is a deep priority and commitment. Regenerative agriculture is not only about improving crop yields and reducing harmful ecological impacts. It is about bringing new vigor to biogeoecological systems while enlivening us as humans.

The legacy of the Green Revolution has been the destructive use of industrial techniques and "miracle technologies"—pesticides, fertilizers, genetically modified seeds, monoculture crops—to maximize yields. Soil and other natural systems are not treated as alive but as machines, essentially dead resources. In the Great Regeneration envisioned by Cox, technology plays a significantly different role. Instead of deploying powerful, poorly tested tools that often shatter the dense, symbiotic web of life in a landscape, the Great Regeneration sketches an agricultural future that revives aliveness through the skillful blending of open-source technologies, ecological wisdom, and local empowerment.

In his seminal history, *Two Bits: The Cultural Significance of Free Software*, Christopher M. Kelty notes how free software (the politically minded precursor to open-source software) is "a kind of collective technical experimental system." It blends conventional practice with daring experimentation to address evolving, real needs. It privileges creative, pragmatic solutions over proprietary business models, entrenched political interests, and even law itself. (Free and open-source software became possible only through clever, elegant "hacks" of copyright law. Expanding the scope and support of law is a part of this new future as well.)

Open-source systems are at once powerful and flexible because they honor individual creativity that can be collectively shared and constantly improved upon. The technologies avoid bureaucratic and political stagnation by privileging the freedom of bottom-up agents over centralized control. They authorize and support creative modification and agile innovation. The focus is not on beggar-thy-neighbor competition and market success that tends toward economic consolidation; it is about cooperative stewardship of dispersed, autonomous systems on a holistic scale. Everyone can flourish together. Instead of intensifying the winner-take-all ethic that often prevails in capitalist markets, regenerative agriculture can deliver maximum effectiveness at low cost. Its "secrets" are democratic participation, sharing and collaboration, transparency and accountability, flexible innovation, and the freedom to localize solutions.

These affordances, and this ethic, are precisely what contemporary agriculture will need to navigate the difficult years ahead. As technology comes to support natural systems rather than disrupt them—through monitoring sensors, software apps, data analytics, networked cooperation, and more—Cox astutely sees a new "silicon-based nervous system" helping farmers to monitor and improve the carbon-based ecosystems of life. Open-source technology can enhance the search for more symbiotic, ecologically respectful forms of agriculture rather than ignorantly subverting the generativity of natural systems. This infrastructure, artfully knitting together agriculture, ecosystems, and technology, will itself become generative. It will usher in new forms of "cosmo-local production" by inviting a global community of agricultural players to collaborate in developing world-class designs while enabling the production of low-cost physical equipment and infrastructures at local levels.

This compelling vision is not without its complications, however. There are, most notably, tensions between open-source communities and capitalism. The history of Big Tech co-opting or neutering the expansive potential of open-source software is a cautionary story. While there should always be room for value-added proprietary business systems that revolve *around* open-source technologies, dangers arise when technology companies attempt to *capture and dominate* a knowledge commons or other shareable system, whether in a legal or *de facto* sense. Google has been adept at using its market power to become the dominant gatekeeper for public-domain content and certain open-source projects, for example. Apple has leveraged its proprietary iPhone (itself based on research and development financed by US taxpayers) to steer developers to work within its proprietary App Store space.

This history points to an important lesson: Open platforms are commons in only a very thin, fragile sense. One of the great achievements of the technologies behind the internet and the World Wide Web was the establishment of shared protocols that let diverse computing networks interconnect and collaborate. The widespread acceptance of nonproprietary, openly shareable protocols has enabled new types of commons to arise for creative works, scholarship, science, and much else—enough so that easy, no-cost sharing on open platforms is considered a commons.

But this proposition needs closer attention. The conflation of "openness" with the "commons" is misleading. *Openness* implies that a technology or resource is itself, automatically, a commons, a presumption that obscures the fact that a business or group of people (Google, say, or a hacker community) at one point decided how the resource could be legally used. The open/closed binary renders the agency of a community invisible because the access rules for the resource (open or closed) are presented as established facts that somehow are inherent in the resource itself. The open/closed binary also encourages people to presume that making a resource free-for-the-taking is the best, most liberating outcome possible. In fact, absent commons governance, it may actually end up inviting investors and corporations to appropriate the "free," shared resource for their private commercial purposes.

It helps to remember that there are all sorts of choices that a community can make about its shared wealth that are not strictly open or closed in character. The group may choose to make a database available for some purposes but demand payment when outsiders use it. They may wish to allow limited uses of certain designs to trusted colleagues. The open/closed binary fails to name the collective power that a group must exercise in curating and controlling the value it creates (code, information, designs, infrastructure).

This is why seeing the commons as a social system (and not just a resource) is so important. Seeing the commons as a social process of governance and provisioning—*commoning*, the verb form—helps a group recognize its own responsibilities in stewarding and protecting shared wealth. "Open" and "closed" are not the only options.

An open platform or body of content may technically be a commons, but it is a very thin and vulnerable one. A robust commons, by contrast, has participants who actively step up to the responsibilities of peer governance, fair-minded provisioning, and social solidarity. In agricultural knowledge commons, this could entail the intelligent curation of information, development of rules for access and use, penalties for violating rules, the arrangement of secure funding, and so on. A robust commons realizes that it must preemptively prevent capitalist enclosures of its shared wealth and nourish its culture of mutual support.

The creation of new types of agricultural commons represents an enormous and necessary leap forward. It is important to keep in mind, however, that a global peer-to-peer learning community, even if governed through open-source principles, could end up privileging short-term, anthropocentric farming goals over the holistic, long-term needs of an ecosystem. We need the many constructive advances that open-source collaboration on a global scale can yield for agriculture, but might this path serve to homogenize the great diversity of farming cultures around the world? An important challenge is finding ways to honor the pluriverse of local farming cultures as they interact on a common platform (the internet) and traffic in the epistemology of Western science and information technology.

In *Braiding Sweetgrass*, Robin Wall Kimmerer brilliantly explains how Indigenous peoples tend to observe plants and other living systems through a different lens than the inhabitants of capitalist modernity. They see ecosystems through their own distinctive cosmovisions, spiritual beliefs, and intergenerational commitments. "Facts" are not self-evident; they arise and flourish within the framing of a living culture in communion with the Earth. George Washington Carver, the great American biologist who developed crop-rotation methods and novel uses for peanuts and other crops, was a scientist but also a mystic. He declared that his agricultural discoveries came from listening closely, and with respectful awe, to what flowers and other plants had to say.

Mindful of this strain in agricultural history, we must be wary of a techno-solutionism ethic that might marginalize the role of human spirituality and culture in agricultural practices. Western scientists once dismissed the *Subak* irrigation systems of traditional rice farmers in Bali as religious superstition best displaced by "rational," modern techniques. It is now seen that centuries of cultural tradition and religious practices have helped Bali farmers coordinate the timing of planting and harvesting on a community scale, so that scarce water supplies can be effectively allocated and pests kept to a minimum. A broad challenge going forward is to bring the insights of modern science (itself undergoing methodological shifts) into closer conversation with culture. Fortunately, there are constructive models for doing just this, such as the global, open-source network focused on rice agronomy, the System of Rice Intensification.

In the pages that follow, Dorn Cox documents and explains a rich convergence of so many forces that are already making agriculture more regenerative, intelligent, and decentralized. He also opens up fresh spaces for dialogue and collaboration that urgently need to expand. May the ideas in this book find a broad audience of readers—and enterprising, creative farmers around the world—as we enter the turbulence ahead.

DAVID BOLLIER
Amherst, Massachusetts
November 1, 2022

Preface

I wrote this book because the regenerative agriculture movement in its current form has yet to make the industrial model obsolete. If we are to succeed, we must transition to a managed ecosystem approach that goes beyond the soil health story that has dominated the sustainability and regenerative agriculture movement in recent years. The story of soil health, often told in the form of a solitary but inspiring farmer or rancher, focuses on the many benefits that accrue when soil is treated as a living, biological resource rather than simply a chemical medium for holding plants upright. These narratives of improving soil health have laid a strong foundation for a regenerative approach, but the solutions we need now involve a broader public narrative about our relationship to science, to each other, and to our institutions.

The new narrative is not about making a false choice to prioritize individual farm transformation, or large-scale public finance and incentives, or relying on powerful private market forces. Rather, each of these things is necessary, but none are sufficient on their own. The new narrative frames these components as branches of the same tree which share a common trunk of global, peer-to-peer distributed learning community. The same tools used to visualize and identify the global instability in our climate, our communities, and economies can now be used to accelerate regenerative solutions that build upon the strength of individual creativity, distributed governance, and marketplace incentives.

In this context an equation emerges: To grow our commonwealth of nature we must add to our commonwealth of knowledge. In the pages to follow, I will explain this equation, which is both new and ancient. This book is about carving out another way, one that embraces the principles of regenerative agriculture as well as democratized, open-source technology, which can disseminate high-quality information, not just to farmers and ranchers, but to all of us as we take on the role of ecosystem stewards. Call it regenerative agriculture meets agrarian public science. Call it the Great Regeneration.

In the 1970s, my parents started a small organic homestead down the road from what was then a sleepy University of New Hampshire in Durham. They were part of a back-to-the-land movement that challenged status-quo assumptions about industrial capitalism and the agricultural orthodoxy of the time. In many ways, my parents, who both had agriculture degrees from Cornell University, were rejecting the tools and tenets at the center of the Green Revolution, a set of initiatives in the 1950s and 1960s that aimed to end world hunger by drastically increasing food production worldwide through the use of synthetic fertilizers and the mono-cropping of high-yield cereal grains. In essence, the Green Revolution was a rejection of small farmers and Indigenous knowledge in favor of corporate monopolies and top-down solutions. Earl Butz, the US Secretary of Agriculture under Presidents Nixon and Ford, summed it up this way: "Get big or get out." A lesser-known quote of his might be even more telling: "Food is a weapon."

My parents could see, like so many before them, that an alternative to being dependent upon an industrial supply chain and commodity agriculture was to adopt, adapt, and build on biological methods of agriculture. In addition to eliminating chemicals and purchased inputs on the farm, that alternative often meant using older, small-scale technologies combined with traditional methods. It meant rejecting the use of large-scale industrial approaches and dependencies. Growing up, I developed an appreciation for approaches that nourished soil, incorporated livestock and perennial plants, and produced healthy and abundant food that you could pick and eat right on the farm; and I was motivated by a promise of economic independence within the local community.

So how did I get from there to here? Why would a child of back-to-the-land homesteaders write a book asking readers to consider the role that transformative technology could play in scaling regenerative agriculture all over the globe? Isn't technology at the heart of the problem? To be clear, I believe it's a misconception that the back-to-land movement was entirely anti-technology. In my experience even the most radical and anarchic of small farmers are inventors, systems thinkers, and technicians of the highest degree. My parents looked to agrarian communities such as the Shakers, Mennonites, and Amish for inspiration; and they were deliberate when it came to the tools they chose to pick up. The key questions of the movement were: Does this tool support or undermine independence and community? Can I make it or fix it within the community where I live and work? What tools and technologies increase efficiency but don't burden the farmer with debt or dependencies? While my parents were asking these questions about their own land, the same questions were simultaneously being asked at a different scale in many postcolonial countries, like India, as they planned for independence, reconstruction, and development following World War II.

The *Whole Earth Catalog* was the bible of both back-to-the-landers and nascent technologists like me, and I remember my parents' living room bookshelf having not just one copy, but each edition. There were other inspiring titles on that same shelf—titles that tell a story: *Reconstruction by Way of the Soil*; *Farmers of Forty Centuries*; *Small Is Beautiful*; *How to Get Out of the Rat Race and Live on $10 a Month*; *A Pattern Language*; *A Long, Deep Furrow*; *Humanistic Economics*; *Malabar Farm*; *Pale Blue Dot*, and many more. The *Whole Earth Catalog*, founded in 1968, was filled with information on ecology, small-scale agriculture, thoughtful technologies, and bold new ideas that could help homesteaders and others establish their independence. With articles ranging from desktop publishing to goat husbandry, the magazine inspired everyone from technology visionaries like Steve Jobs to pioneering organic farmers. In a 2013 article in the *Guardian*, technology writer John Markoff is quoted as saying that the *Whole Earth Catalog* was "the internet before the internet. It was the book of the future. It was a web in newsprint." It was no accident that the first edition of the *Catalog* featured an early NASA satellite photo of the sphere of the Earth. The significance of this image, a symbol of our shared destiny, will become apparent in Chapter 1, where we will learn that the *Catalog* was part of an older and ongoing human ambition to document and share knowledge.

Stewart Brand, the *Catalog*'s founder and a techno-utopian visionary, bluntly summed up how I feel about the role of technology and human ingenuity when he said: "We are as gods and might as well get good at it." In my view, we've yet to truly "get good at it." Historically, we've either denied our power or we've abused it.

Before we take a deep dive into this new world of digital interconnectedness and ecological regeneration, I want to return to my childhood. Although I didn't know it at the time, my family farm was setting the stage for the work I'm doing today to develop collaborative and open-source digital platforms that will bridge the knowledge gap for people creating healthy soil and thriving regional economies and ecosystems. For me, growing up on a homestead wasn't especially countercultural— it was just normal. I didn't feel sheltered or cut off from the rest of the world. Far from it. I remember my grandmother regaling me with stories about pre-Communist China, where her father served as a medical missionary and narrowly escaped down the Yangtze River to Shanghai as their home near Changsha was bombed by the Japanese. I grew up with family friends who had been some of the first Westerners to visit Tibet and were close friends with a mountaineering network in Nepal and Alaska. I remember talks about fusion and energy density with my grandfather, who worked developing experimental equipment in his home physics lab years after leaving the University of California, Berkeley. Thanks to these family influences, I felt not only a deep attachment to the farm but also a connection to world events and scientific achievements, even as my siblings and I learned to use a shaving horse or carve a spoon from birch root.

I was identified by my peers in school as a farm kid, but my experience didn't fit that stereotype. My introduction to technology's liberating potential emerged as I entered sixth grade and was struggling with severe dyslexia. Spelling and writing were a painful barrier to expressing my knowledge, even as I was able to draw and illustrate and sculpt in three dimensions fluidly. I credit an early assessment, which advised my mother to "get this kid a computer," for changing my trajectory. It was through typing and spell-checking that I was slowly liberated from my disability, which despite all odds led to increasingly academic pursuits that continued to mix technology, international context, and my agricultural experience.

Following in my parents' footsteps, I went to Cornell University, where I studied international agriculture and rural development, and often walked across the bridge overlooking Carl Sagan's home, jutting out over a waterfall spilling into a deep gorge. The place became a special touchstone for me at the time, but little did I know the significance it would gain a decade later, as Sagan's writing influenced my thinking about public science, soils, and climate and life on Earth. With my homesteading and farming background, I questioned what I was being taught about export-oriented agricultural development theory, including the role of technology and mechanization. This kind of economic development had too many negative environmental and social side effects, including soil loss, decreased weather resilience, debt, displacement, and migration— what are sometimes called *externalities* in classical economics. My experience on the family farm taught me that social and ecological effects were, in fact, *internal* to everything we did. I was shocked when, in a final course in international development economics, my professor presented evidence suggesting that much of the theory that we had just been taught on large-scale agriculture was not only *wrong* but *counterproductive*. I will never forget his admonition that we now had to start over. "Good luck," he told us.

Of course, we did not need to "start over" but rather build on centuries of previous knowledge that had been obscured by the rush of industrialization that characterized the 1970s. It became clear to me that the next generation had to do better.

Before I embarked on a new path of challenging conventional assumptions and eventually returned to the family farm, I went on a winding, multiyear detour that took me all over the world, from Wall Street to Argentina to Hong Kong and into the worlds of high finance, internet start-ups, and data visualization and analysis.

While still in college, I had worked with a small venture group associated with MIT to develop a data visualization tool called Global Watch, a crude precursor to Google Earth. The hosting and network infrastructure wasn't quite mature enough to support the vision at the time, so after I graduated from Cornell, rather than pursue the technology further or return to the family farm, I followed a group of friends to New York City for what turned out to be a brief stint in equity research and analysis at a renowned investment bank. I wanted to link my agrarian experience and education to global finance and to understand the levers of power that were driving development. I quickly became disillusioned, however, by the vast disconnect between energy, land, people, and knowledge. Even at one of the world's largest banks, where the highest-quality data should be available, consequential decisions were being made on limited, coarse, and disconnected information. There was a great deal of market data, but almost no environmental data that tied decisions to the natural world. I needed to find another path.

The next leg of my journey took me to Buenos Aires, where I was inspired by an invigorating, cosmopolitan, agrarian culture. The Argentines I met and worked with, even those from posh neighborhoods, seemed connected to soil and agriculture, which they considered to be the foundation of their economy.

Returning from Buenos Aires, I reconnected with friends in New York and together we launched a tech incubator that rocketed me to Hong Kong and around Asia in a project that quickly grew from a small core team to more than 200 people in less than nine months. It was the peak of the Asia tech bubble, and despite the excitement of being in the middle of fast-moving developments, I felt unmoored. I found myself taking ferries from Hong Kong to nearby islands and hiking to remote jungle beaches on the China border just to reconnect with the natural world. My disillusionment grew. I wanted to ground technology and enterprise in the real world, not live in the abstractions and aberrations of financial modeling spreadsheets with missing variables that seemed to be driving decision-making.

While this globe-trotting was an exciting and formative time for me, I missed home deeply. I longed to ground the technological innovation I was discovering in the early 1990s and bring it back to the family farm. So I returned home to New Hampshire. It was time to rethink my future. Within a few years, I met my wife and returned to farming, while still remaining active in the world of tech start-ups and software development. It felt like I was starting to find my calling. It wasn't in the farthest reaches of Asia, or on Wall Street or Silicon Valley—it was right back at home in the lush green hills and temperate forests of my childhood.

My wife and I joined a young farmers group. The discussions among the members enabled me to combine local agriculture with academics, a nexus with so much potential for farmers looking to free themselves from the industrial mindset. As farmers, we were told to squeeze every cent from a product but never to challenge the system itself. President John F. Kennedy summed up the dimensions of the problem when he said, "The farmer is the only man in our economy who buys everything at retail, sells everything at wholesale, and pays the freight both ways." Within a year of returning home, I dug in deeper to my agricultural convictions by helping start both the Great Bay Grain Cooperative and the New England chapter of the National Farmers Union, where I served as a founding board member. At that point, however, our focus was on local production, regional equity, and agriculture-based energy independence—not on the more ambitious and holistic goals of regenerative agriculture that were yet to come.

It was at a Farm Bureau meeting in Upstate New York that my chance encounter with regenerative farmer extraordinaire Klaas Martens changed everything. I saw him speak with pride and conviction about his grain farming operation and how he was using cover crops and reduced tillage to improve his soils. He spoke about making a business where he and his wife, Mary-Howell, were able to transform the local economy by marketing directly to customers and rebuilding a local grain processing facility to supply growing demand for organic grains and cover crops. They had simply looked at the situation, made an assessment, and decided to change how they approached both business and agriculture. On that day I felt like he was speaking directly to me, and it marked the beginning of a new phase.

Little did I know how important that moment would be, or what it foreshadowed, when I again met up with Klaas at a launch event for Soil Health Institute nine years later. He provided inspiration and an early introduction into what would become the current regenerative agricultural movement in the United States. I started reading up on the subject as if my life depended on it, because it felt like all life did depend upon it. I started growing grains, cover crops, oil seeds, and grazing animals. I started shifting my focus from technology development to transforming available technology and finding appropriately scaled equipment worldwide that would help lower costs and bring more productive capacity and resilience to our own local food system. It was clear at that point that I was not ever going back to business as usual.

I got even more excited about linking agriculture and energy systems, initiating the Oyster River Biofuel Initiative, which aimed to create local biofuel systems based on crops such as sunflowers, which I had seen grown in Argentina as a feed and fuel crop. A few years later, that effort led to the launch of a nonprofit organization called GreenStart. The group had the goal of fostering innovation for a resilient energy and food systems for New Hampshire by providing technical education and practical agricultural examples. We also had the aim of supporting collaboration between agricultural extension providers, researchers, the US Department of Agriculture (USDA), and agricultural and conservation organizations.

Around 2002, a number of pieces came into focus when I linked up with San Francisco–based open-source hardware communities through the internet forum Biodiesel Now. The open-source culture was full of inspiring, charismatic leaders, including countercultural Google dropouts and people such as Maria "Girl Mark" Alovert, who might be described as a Bay Area cyberpunk. She was well known for the photocopied *Biodiesel Homebrew Guide* she published at cost, which included designs and instructions for the Appleseed biodiesel processor that sparked a national open-hardware, DIY biodiesel movement. I was beginning to see the power of linking open-source tools with agricultural knowledge and local manufacturing culture. The *Whole Earth Catalog* had published its last issue many years before, but its DIY spirit remained.

In my hunger for exploring ways to create more value from the land and to challenge assumptions about scales of production, I was inspired again by what I had witnessed in Argentina years earlier. I had seen on a single farm that pelletized feed could be produced from sunflowers grown on a small portion of cultivated land and substituted for the diesel fuel needed to run all the equipment required for the operation. In this simple setup, I saw a substitute for a global supply chain dependency that could be owned and operated at the level of a single farm or a small cooperative. How could it be that a bit of the right knowledge and some local manufacturing capacity could substitute for a multibillion-dollar global diesel supply chain? Could a local system deliver a gallon of fuel for equal to, or less than, its fossil fuel equivalent, even without a subsidy, and produce a high-quality culinary product and animal feed at the same time? Could large-scale industrial processes at the core of our economy be reduced in scale to democratize access to energy? Why was this system so hard to discover? What was the limitation? The questions were intoxicating.

The answers, I realized, were found in accessing knowledge and breaking down the barriers to sharing technical insights. The status quo of industrial knowledge being protected to extract the monetary value of a competitive advantage was flipped on its head. I began to see what is possible when people have new ways of sharing knowledge. However, even with the powerful example of informal exchange of technical knowledge over internet forums, the technology was not yet mature enough to meet the urgency of energy security, climate change, and agricultural transformation.

This realization propelled me into a PhD program at the University of New Hampshire. It also led to the creation of Farm Hack, an online forum and global community aiming to improve how we share technical agricultural knowledge. I was motivated by the belief that the democratization of useful tools and techniques forms the foundation of regeneration, agriculture, equity, and inclusion. In 2011, I linked up with the National Young Farmers Coalition and learned about their work with the new MIT D-Lab, an international sustainable development lab with a social equity focus. The D-Lab organized the first Farm Hack to bring together designers, farmers, engineers, and others to share the results of their work.

That same year, I connected with Courtney White, who was leading the Quivira Coalition and writing a book titled *Grass, Soil, Hope: A Journey Through Carbon Country*. We met at my family's farm and sat for hours around the picnic table, where a beautiful conversation emerged about farmers, ranchers, environmentalists, and scientists finding common cause—a conversation that is still unfolding in this book. I realized at that moment, in talking with Courtney, that I was far from alone in my journey to assemble the pieces of a potentially world-changing view of humans and nature—a view that fused accessible technology and the free distribution of agricultural knowledge.

After decades of practice-based innovation on farms and ranches, as well as the wisdom we've gleaned from Indigenous practices and anthropological records, we know what resilient agriculture looks like. Recently, we began to understand the critical roles carbon and carbon cycling play in agriculture as well as mitigating the climate crisis. However, very few people understand the boost that new low-cost observational technology—that is, silicon-based systems—can give carbon-based agroecological systems, and few have yet to see the hopeful role that open source can play in data-sharing, networking, and telling powerful stories about ourselves and our environment. This is not just a story about regenerative agriculture and the planned obsolescence of industry-caused climate change, but of a longer human endeavor to grow both our commonwealth of knowledge in concert with our commonwealth of nature.

Introduction

We believe in change and we believe in it transpires when we need it.
But the ships that we stand are not see fit.
It's these vessels filled with the chaos of commerce that are leading us
    into the wreckage but I swear we're at the helm.
We've got the tillers in hand
And the truth that was lost we can still understand.
We need to build bridges over this splintered land
Before the hour glass cracks and spills it's sand.

        — KAE TEMPEST, "Parables" (2012)

We live in an extraordinary era in human history. Never have we had more influence, both positive and negative, over our environment and our future. Back in 2008, Tom Atlee, a long-time peace and ecology activist, wrote "I've come to believe that things are getting better and better and worse and worse, faster and faster, simultaneously."[1] Today, I would add, things are also getting more and more complex. We can communicate more directly with each other than ever before through vast systems of technology. We can observe nature and the cosmos, and contemplate our unique position in the solar system, with far more accuracy than ever before, revealing spectacular details of interconnectedness and knowledge, and yet we are simultaneously presiding over the sixth mass extinction of life on Earth. Algorithms help create financial instruments of extraordinary complexity and at the same time help us understand the damaging consequences of climate change. In her book about mass biological extinction, *The Sixth Extinction*, journalist Elizabeth Kolbert comments on this very human paradox, writing that we are "the sort of creature that could wipe out its nearest relative, then dig up its bones and reassemble its genome."[2]

Technology is more and more ubiquitous, too: Cameras and GPS units are now in every pocket; environmental sensors are embedded in handheld devices; satellite monitoring is now available at low cost; and an emerging network of devices, often called the Internet of Things, allows for global data sharing in real time. The accelerating rate of processing power in microchips, described by Moore's Law, combined with huge technical advances in data storage capacity and batteries, means we can connect and share knowledge with each other as never before. Much of the infrastructure was first harnessed, often without our knowledge, as "Big Data" analysis driven by business, government, entertainment, individual health, and security interests. The first generation of information technology centralized power and wealth rather than creating greater independence for citizens or for collaboratively solving environmental challenges. The result is a huge loss of potential. Not long ago, the high cost of this technology limited its distribution to governments or large corporations. But costs have fallen to a point where this technology is accessible to nearly all. For less than five dollars, it is now possible to purchase an open-source UNIX computer with Wi-Fi and Bluetooth connectivity contained in a device that is smaller than the palm of your hand. Memory cards smaller than a fingernail can hold all of the core documents of human history and scientific knowledge. High-definition digital cameras that can be purchased for a few dollars are now embedded in circuit boards on phones.

Global communication is now possible at a tiny fraction of the earlier costs while using comparatively fewer materials and less energy to create the same functional infrastructure. Just twenty years ago, while living in Argentina, I had the choice to go to a pay phone booth to make an international call at a dollar per minute or go around the corner to an internet café where I could browse online for ten dollars an hour. The internet, a testament to human collaborative visions of both utopia and dystopia, achieved the density of connections on par with the neural complexity of a human brain in less than fifteen years. We are now just beginning to harness the potential for global collaboration as a means of unlocking the mysteries of evolution and the diversity essential to sustaining life on Earth. This collaboration is critical to restoring the regenerative capacity of nature, even as the same technology is being harnessed to accelerate its degradation. A race is on, in other words, and we have the opportunity to choose which way to go. Much of this technological transformation has taken place within the last decade. The time is ripe to harness our distributed powers of observation and create a global knowledge commons that makes scientific, and particularly agricultural, knowledge accessible to everyone everywhere. As we will discover throughout this book, this is not a new idea. The first known example of agricultural knowledge being published and shared took place nearly 4,000 years ago in Mesopotamia, where archaeologists have unearthed clay tablets inscribed with advice on watering crops.[3]

The goal of knowledge commons is the transformative *use* of our knowledge in service of a specific action—namely, regeneration. In other words, repairing, maintaining, and expanding life itself. Regeneration is the opposite of extraction, which we have been conditioned to expect as a price of growth and progress in human societies, and which has been the primary focus of industrial technology. But not all technology is industrial, and industry does not have any special claim to all technology. As a process, regeneration does have energy and infrastructure costs that must be extracted from nature or that depend on industrial technology, but with regeneration it is ultimately more important to emphasize how those resources may be optimally used and recycled in service of naturally productive systems.

In evolution, regeneration means constant revision to genetic codes in response to an ever-changing environment over millions, even billions, of years. Even after mass extinction events, life on this planet has found ways to recover, grow, and thrive anew. On a farm, regeneration means harnessing these same evolutionary forces, only over the span of a few human generations rather than eons. The term *regenerative agriculture* has experienced a meteoric rise in public interest and is being widely used in nonprofit and corporate circles alongside terms such as *climate-smart agriculture* and *natural climate solutions*. This explosion of excitement and engagement speaks to the power and potential of these concepts and practices, but it also risks being co-opted and simplified into "regenerative practices," which are often used interchangeably with techniques that build soil health, such as cover crops, minimal tilling, plant diversity, and composting.

Regeneration as a concept is more powerful than a set of practices, or even principles. Regeneration is not a practice but a *process*—of observation, planning, collaboration, and action—with outcomes that can be measured in improved ecological function over time. Regenerative agriculture restores and maintains soil health and fertility, protects watersheds, supports ecological and cultural diversity, and expands economic resilience. Soil health is the ability of soil to function biologically, chemically, and physically in order to sustain productivity and maintain environmental quality, as well as to promote plant, animal, and human health. Regenerative agriculture focuses on creating the conditions for life above and below ground, and takes its cues from nature, which—let us not forget—has a long record of successfully growing things and creating a livable planet. By recarbonizing soils via photosynthesis and biology, particularly on degraded land, regenerative agriculture can sequester additional atmospheric carbon dioxide, making it a key part of the solution to climate change.

We have been given the gift of great libraries of genetic code and time- and nature-tested processes for generating life upon which we can grow food, accelerate the healing process if land has been damaged, and adapt to changing local conditions, including those linked to climate change. Although the genetic library available to us is just a fraction of the diversity that has existed on Earth over its history, it forms a rich source code library for regeneration upon which we can build and grow productive, verdant landscapes. This is just one reason why seed banks and healthy habitats that support genetic diversity are so important.

I have participated in the process of adapting varieties of hard winter wheat to New England growing conditions, which has within a decade created new local grain economies. I have seen the power and economic benefit of being able to save our own cover-crop seed year after year. I have planted open-pollinated flint corn varieties from seeds that my father had saved each year. I have watched seed originally stewarded and saved by the Narragansett tribe of Rhode Island outperform commercial hybrids. The acts of observing, selecting, and sharing are at the heart of the process of regeneration, creating the foundations for repairing and creating (or re-creating) a land of milk and honey within our lifetimes.

This goal is different from that of sustainability, a term introduced into the popular lexicon in the late 1980s across wide sectors of society, including Wall Street. *Sustainability* is hard to define, and perhaps its elusiveness is a reason for its popularity. In theory, sustainability means not compromising the future with actions taken today. In reality, sustainability often translates into maintaining current conditions and lifestyles well into the future, even if the practices being used are the ones that put us into jeopardy in the first place. Take degraded farmland, for example. Often a consequence of industrial farming practices, degraded lands are prevalent around the world. As Gabe Brown, a farmer and regenerative-agriculture pioneer in North Dakota, puts it in his book *Dirt to Soil*, "Everybody wants to be sustainable. My question is why in the world would we want to sustain a degraded resource? We need to work on regenerating our soils, not sustaining them."[4]

Brown transformed his family's conventionally managed, eroded, and depleted land into a biologically rich, healthy, and productive farming operation by turning *dirt* into *soil* with regenerative agriculture. "When we purchased the farm in 1991, the infiltration rate on our cropland was only a half-inch per hour," wrote Brown. "That meant when a big storm came along, dumping two or three inches of rain, most of the water left the farm in a hurry, usually taking a bunch of topsoil with it." After raising the carbon content of his soil from 2 to 5 percent (which is a lot) by using agroecological practices such as reduced tillage, grazing, and cover crops, this situation changed dramatically. "By 2009, the infiltration rate had risen to more than ten inches per hour thanks to well-aggregated soils due to mycorrhizal fungi and soil biology."

Brown likes to joke that he could never go fishing because there were no earthworms on his farm under conventional management. When he stopped using chemicals, however, earthworms appeared! The soil was alive and could support abundant life beyond the crop. It was a sign the land had begun to heal.

Dirt is mineralogy and chemistry—individual minerals and elements, including calcium, phosphorus, and potassium—but it's not life. Soil is biology—minerals structured through respiration, reproduction, growth and death of plant roots, bacteria, fungi, protozoa, nematodes, and earthworms that are part of a soil food web. Life, in other words. Getting plants to grow in dirt is chiefly a matter of getting the chemistry right and applying it according to a calculated prescription. Getting plants to grow in soil, by contrast, means getting the biology right. If the land is devoid of life, you need to foster the conditions for the return of living organisms. Worms are as important to soil life as blue whales are to ocean life, but the real key to regenerating soil is represented in the living carbon cycle. Based on the key ingredients of sunlight, water, nutrients, and soil microbes, the process by which atmospheric carbon dioxide gets converted into soil carbon and produces an oxygenated atmosphere through living organisms has been going on for more than a billion years.

Brown, like others, followed a personal journey of discovery by observing the outcomes of practices he implemented, which improved his soil's health over time, including planting diverse cover crop mixes year-round instead of tilling. Brown's personal journey is also part of the human history marked out on the landscape—the history we can now see through real-time satellite imagery. Agricultural success stories like Brown's are a source of inspiration, and at the center of this book's mission is a call for us to all take the next several steps. The mission is to see all of our stories as pieces of a developing, global pattern that can renew landscapes and enable us to not just adapt but to thrive as we mitigate and adapt to climate change.

Reversing Degradation

Even with increasingly sophisticated approaches to regenerative agriculture, as Gabe Brown discovered, there's an important requirement for successfully regenerating degraded soil: *We must reduce tillage in most soils in most production systems.* Disturbing the soil every year by ploughing breaks down the building blocks that make up the living spaces supporting complex biological communities and exposes the unprotected soil to wind and water erosion while releasing into the atmosphere carbon that was previously stored as part of the biological metropolis underground. Tilling breaks up soil aggregates (tiny clumps of soil glued together to form spongelike pores), destroys critical fungal networks, and can result in bare or compacted soil that creates a hostile environment for microbes, earthworms, and other forms of life. And when life departs, erosion soon follows.

Soil erosion is the most widespread type of land degradation in the world. In nature, land is generally buffered against the erosive effects of wind, rain, and snowmelt by ground cover, such as grass, shrubs, trees, and even weeds. When this vegetative cover is disturbed or removed, the soil is exposed. A single storm can turn a farm furrow into a small gully. The application of chemical fertilizers and pesticides to crops, which can damage the microbial life and nutrient cycling that glue soil particles together underground, can also cause soil degradation. As soil leaves the farm, it carries essential nutrients for plants along with it. Soil loss, in combination with climate change, is predicted to reduce crop yields by 10 percent globally in coming decades and by as much as 50 percent in vulnerable areas. This loss has already led to significant human migration, increasing the risk of conflict and political instability. Eroding soil takes carbon with it, as well. Croplands have lost between 20 and 60 percent of their original soil carbon since cultivation began, and land under conventional agriculture continues to be a source of greenhouse gasses. According to one study, since 1850 about 35 percent of all human-generated emissions of carbon dioxide have originated from land as a combined effect of degradation and land-use change, such as forest clearing.[5]

Land degradation has been happening for more than 10,000 years, ever since diverse stewarded perennial plants began to be displaced by cultivated annual crops as part of the agricultural revolution. Grazing animals were domesticated starting more than 8,000 years ago, and plows were in use at least 5,000 years ago. The vast forests of Europe were largely gone by 1000 BCE. Populations grew, requiring greater amounts of food and fiber, which increased pressure on the land. As the ancient Sumerians, Greeks, Romans, Mayans, and Chinese could tell you, soil erosion matters. Geologist David Montgomery noted that, time and again over the course of human history, social and political conflicts are exacerbated when there are more people to feed than can be supported by the land. Civilizations don't disappear overnight, and they don't choose to fail, he noted. More often they falter and then decline as their soil washes away. Rome didn't so much collapse as crumble over generations, as agrarian traditions and yeomanry of the republic were replaced by large slave-holding estates of the empire, which stunted crop yields over time, wrote Montgomery, as erosion and soil degradation steadily sapped the land's food-growing capacity.[6]

In North America, millions of tons of topsoil are eroded annually from farm fields into the Mississippi River and the Gulf of Mexico. America's farms lose enough soil every year to fill a pickup truck for every family in the country. It is a similar situation globally. According to the United Nations, between 25 and 30 percent of all ice-free land on Earth exists in a degraded condition, affecting three billion people who depend on this land for their livelihoods.[7] Most of the land degradation is the consequence of industrial incentives that drive shortsighted agricultural practices and deforestation, resulting in twenty-four billion tons of fertile soil being lost each year to wind and water erosion. If this trend continues, the percentage of degraded land will jump significantly by 2050—the same year the global population is projected to increase by a third, to ten billion people. This extra demand for food, feed, fiber, and fuel—in combination with environmental and social stresses caused by climate change and the loss of biodiversity—highlights the grand challenge and opportunity. The choice is stark. Will the next, larger generation of people result in even more stress on the systems that support life on Earth, or will they contribute to restoration, regeneration, and abundance through improved stewardship?

When my parents started their organic farm, soil erosion wasn't on their minds. The organic movement brought with it a lot of agrarian idealism, expressed by back-to-the-landers like my parents, who were as interested in the independence of smaller-scale agriculture and community-building as they were worried about reliance on chemical inputs. Early organic farmers stood in opposition to the trend toward larger-scale agriculture and a focus on low-cost commodities to "feed the world"—the get-big-or-get-out philosophy mentioned earlier. Regeneration expands upon decades of organic research market and infrastructure development. Today, with the climate crisis and other global issues, including food vulnerability due to extreme weather, the cultural context of what we want and need from agriculture is much greater. Farmers and ranchers and all land stewards are now asked to not just feed the world but also steward the environment.

That's a tall order. What we need to accomplish it is a global knowledge commons that democratizes access to trusted scientific information that supports growing food; the knowledge to mitigate flooding, drought, and weather extremes; the preservation of biodiversity; and the restoration of ecological functions to our landscapes. More important, much of what we are now asking of agriculture is not just food, fiber, and fuel, but also environmental outcomes that are invisible to us without data-driven stories and people to document and share them. The scope and scale of our shared endeavor requires us to create cultural and economic links between population centers in cities with the rural landscapes that provide food, water, climate regulation, and habitat; and, furthermore, it requires that we democratize access to our common library of biodiversity for all.

Regeneration can do this. Regenerative agriculture, and the data-driven adaptive management that it requires, shifts our shared conversation from processes to outcomes. As data comes in—on changing carbon levels or water infiltration, for example—a land steward can adapt their management practices, improving outcomes over time. The challenge in using the adaptive management approach lies in finding the correct balance between gaining knowledge to improve management in the future and achieving the best short-term outcome based on current knowledge. Regeneration leads us on a path toward a collaborative marketplace of both goods and environmental services that rewards outcomes, establishes incentives for improvement over time, and creates abundance and health as a byproduct. Verification and validation of environmental change, difficult to achieve in detail as recently as a decade ago, has become much less expensive and much easier to obtain and share today because of rapidly advancing, low-cost, and open-source technology. This change is making regenerative agriculture a practical and profitable alternative that can be used across all scales of production systems. The principles of regeneration and restoring soils and biodiversity apply everywhere and can form an ever-evolving, sharable global recipe book that can be adapted and adopted locally to address everything from caring for a window box garden on a fire escape, growing food in refugee camps, and growing windbreaks to planting hedgerows along grain fields, stewarding grasslands across millions of acres, and managing orchards and vineyards.

Abundance

A key to regeneration is the theory of abundance. Mainstream macro- and microeconomics taught in schools for business purposes—the kind taught to me in undergraduate and graduate programs—does not take into consideration the relationships involved with living landscapes. These are theories for a world focused on efficient extraction, allocation, distribution, and competition for scarce resources. However, many of the core functions of regeneration—the foundation of nature itself—is based on environmental goods that are highly abundant and whose richness grows as the system improves, creating a positive feedback loop, a virtuous circle that improves with time. Sunlight, for example, provides energy for photosynthesis in plants and is the key to their growth and health. Not only is sunlight free, but it is also extraordinarily abundant and distributed globally. Despite being based on natural abundance of biodiversity and nature's propensity to reproduce itself, the history of agriculture, with few exceptions, has been one of extraction, where disturbed soil degrades over time, making it more and more difficult to grow and sustain a crop. In contrast, regenerative agricultural reverses this pattern. Rather than degrading, the central premise of regenerative systems is to improve soils, such that they hold more water over time and thus become more resilient to floods and droughts. This powerful phenomenon is something that I witnessed firsthand during the first five years of managing fields on my family's farm, as we began to change the way we approached soil management.

Similarly, a regenerative economy is based not on the depreciation of assets but the appreciation of assets through use, like the strengthening of a muscle. Correct use, but not overuse, of a muscle does not diminish it but increases strength over time. The agricultural knowledge commons is another example of a "nonrival good"—something whose value does not diminish or depreciate, but improves over time with use. Examples include Wikipedia and the often invisible open-source software infrastructure that forms the backbone of the internet we all use, even when what we interact with daily are often commercial interfaces. It is in this context that the complexity and redundancy of biological systems are useful guides in that their strength and resilience and continual improvement are reflected in our shared commonwealth of nature. This interconnectedness is the foundation of an abundant and regenerative economy. My farm's use of sunlight, rainwater, a local corn variety, no-till cultivation practices, or the sharing of specific equipment designs do not diminish others' use of the resource—rather it has the potential to improve it over time. The abundance of these shared resources makes all the difference.

As the scale of land degradation has grown in recent years, many land stewards have shifted their attention beyond sustainability and toward the goals of restoration and regeneration. Buoyed by a renewed interest in traditional Indigenous practices, on-the-ground innovators—often led by farmers, ranchers, and other land stewards—are incorporating observational and exploratory scientific processes into their own operations, enabling them to continually adapt and improve outcomes over time. If these observations can be recorded and shared, every farm, backyard, rooftop, or windowsill can turn into a data-driven applied research farm. Adaptive management was a new term to me when I first came back to the family farm, but as I was introduced to soil health measurements, I quickly saw the power of adaptive management as an alternative to conventional practices and as a process that empowers land stewards to approach their land as a living and constantly changing thing. As a consequence of this work, we all now have a constantly expanding toolbox of observation and land-restoring practices to help us increase our adaptive capacity.

The common thread is *what actually works* in practice in the natural world. "I follow five principles that were developed by nature, over eons of time," Gabe Brown wrote in his book. "They are the same any place in the world where the sun shines and plants grow. Gardeners, farmers, and ranchers around the world are using these principles to grow nutrient-rich, deep topsoil with healthy watersheds." Here are Brown's five principles for regeneration, which are also now echoed in the US Department of Agriculture's own description of soil health:

LIMIT MECHANICAL, CHEMICAL, AND PHYSICAL DISTURBANCE OF THE SOIL. Tillage destroys soil structure. It is constantly tearing apart the "house" that nature builds to protect the living organisms in the soil that create natural fertility.

KEEP THE SOIL COVERED AT ALL TIMES. Bare soil is an anomaly—nature always works to cover the soil. Providing a natural "coat of armor" protects the soil from wind and water erosion while providing food and habitat for macro- and microorganisms.

STRIVE FOR DIVERSITY OF BOTH PLANT AND ANIMAL SPECIES. Grasses, forbs, legumes, and shrubs all live and thrive in harmony with each other. Some have shallow roots, some deep. Some are high-carbon, some are low-carbon, some are nitrogen-fixing. Each of them plays a role in maintaining soil health.

MAINTAIN LIVING ROOTS IN THE SOIL AS LONG AS POSSIBLE THROUGHOUT THE YEAR. Living roots are feeding soil biology by providing its basic food source: carbon. This biology, in turn, fuels the nutrient cycle that feeds plants.

INTEGRATE LIVESTOCK TO RECYCLE NUTRIENTS AND INCREASE PLANT DIVERSITY. The grazing of plants stimulates the plants to pump more carbon into the soil. This feeds biology and drives nutrient cycling. A healthy, functioning ecosystem on a farm or ranch must provide a home and habitat not only for farm animals but also for pollinators, predatory insects, earthworms, and other organisms that drive ecosystem function.

The disruptions in climate and global economies now underway provide an opportunity to think much more boldly about abundance. We have for the first time in human history the tools to dramatically change our relationships to each other and with the environment. Through an understanding of a regenerative economy and a new understanding of agriculture, we have the ability to transform our modes of governance. We are no longer limited to organizing large-scale efforts through nation-states or corporations. After decades of practice-based innovation on farms and ranches, we know what

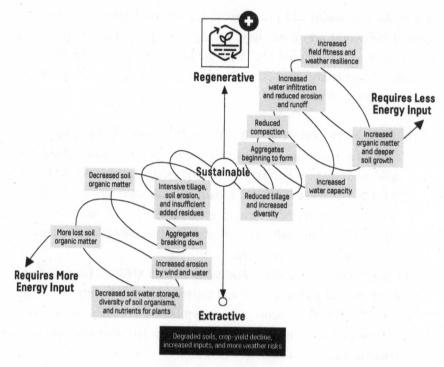

From Extractive to Regenerative Cycles: Growing Healthy Soil. Management decisions are the key variables to growing healthy soil and supporting regeneration over time.

regenerative agriculture looks like on the ground. Recently, we have begun to develop a deeper understanding of the profound roles that carbon and carbon cycling play in agriculture and their potential to mitigate the climate crisis. Low-cost observational technology can document this cycle in great detail.

We are also beginning to explore the critical function that open-source platforms can have in promoting successful agroecological systems through knowledge sharing networks. Combined, these elements have the potential to radically reorganize how we manage food production and natural systems. To meet rising challenges, agriculture needs to transition to a public science that expresses our common understanding of the world. It needs to move away from a solitary rural enterprise to a generation-spanning, shared human endeavor—perhaps one of the largest applied science projects ever attempted. By employing public science and working together, farmers and ranchers can become skilled ecosystem managers, transitioning agriculture from the false inevitability of a "tragedy of the commons" to the possibility of a "triumph of the commons."

I have seen the power of drawing globally from our shared common-wealth of agricultural knowledge. Doing so has greatly benefited my own family, community, and land. Just in my own agricultural journey, I have found small-scale combine harvester and grain knowledge from China, oil-seed-press information from South America, water management and tillage approaches from Australia, cover crop techniques from Canada, and community organizing and market development experience from the South-western United States and rural India.

Throughout this book we will explore examples of how the work of shar-ing knowledge is already underway. These examples are, of course, a small reflection of the vast innovation that happens every day on every farm, and they are intended to highlight the vast potential for sharing across a rich and diverse global network. We each need everyone everywhere to have access to the best possible agricultural knowledge, because together we have the capac-ity to improve soil health faster than we thought possible. At the same time, we must also demand more from agriculture: not only to equitably produce more food, fiber, and fuel, but to provide critical environmental services such as clean air, water, habitat, and biodiversity. Regenerative agriculture has the power to change how we interact with each other and our environment and to provide a pathway to growth and prosperity for the majority of the people on Earth.

Agriculture has the potential to create abundance where there was once scarcity. It has the potential to build bridges across urban and rural divides, and across cultural and language boundaries. It is a multigenerational endeavor rooted in a shared human history, in the joy in observing and man-aging the small details of nature, in the meals we share, and in the landscapes we love. When thinking about agricultural tools and technology in general, it is important to think broadly and explore technology *not* as a product of businesses but as a reflection of our shared values and understanding of our world. How technology is created and curated are reflections of our values, ethics, and choices we make together. Marie Curie famously said, "Noth-ing in life is to be feared, it is only to be understood. Now is the time to understand more, so that we may fear less." It is a time to transform fear into knowledge, and knowledge into hope. It is a time to share knowledge so that we can take informed action together, not retreat into the many "isms" that narrow our paths. Nature cares not at all about our prejudices, which only serve to reduce our potential vision, scale, and pace of collaboration.

We have a choice: allow ourselves to be dominated and controlled by technology or harness it to understand and improve our shared world. We can create new tools that move from *precision* agriculture (large-scale, input-dependent, consolidated, and mechanized systems) to *decision* agriculture—systems that work at all scales, mechanized or not, that are biological, diversified, integrated, democratized, localized, and innovative. Public and participatory science will enable regenerative agriculture and supporting economies to scale, not as a niche but as a new ubiquitous system of management—an elusive goal until now.

For farmers, ranchers, and other land stewards, the power and accessibility of the shared commonwealth of open-source tools can enable more rapid improvements to soil health, as well as payments to draw down atmospheric carbon, improve wildlife habitat, provide clean air and water, mitigate droughts, floods, and fires—all while providing deeply meaningful work to those involved. Land stewards and agricultural communities will benefit by creating broad-reaching alliances across academia, industry, and commerce, unified through open-source research and technology communities. When agriculture becomes a shared endeavor, larger communities can solve complex challenges and help illuminate the intrinsic value we all steward through new markets and ways of valuing our commonwealth of nature beyond the production of food, fiber, and fuel.

The onset of the COVID-19 pandemic in 2020, coupled with extreme storms, wildfires, and flooding around the world, put in stark relief the vulnerability of our current food system and the lack of resilience in our land management practices. It became clear that we are teetering on an edge. The choices we make today will affect the quality of civilization in the not-so-distant future as well as the trajectories of life on Earth. Until now, however, our ability to make a positive impact at a scale that matters has been limited. The connections between a silicon nervous system of sensor and communications technologies and the carbon ecosystem, in combination with knowledge sharing, can change the trajectory dramatically and more quickly than I might have thought possible a decade ago. Individuals, businesses, governments, and communities around the world are waking up to the potential for a Great Regeneration within the next generation.

CHAPTER 1

The Good Anthropocene

At the moment of Sputnik the planet became a global theater in which there are no spectators but only actors.

—Marshall McLuhan

In controversies about technology and society, there is no idea more provocative than the notion that technical things have political qualities. At issue is the claim that the machines, structures, and systems of modern material culture can be accurately judged not only for their contributions of efficiency and productivity, not merely for their positive and negative environmental side effects, but also for the ways in which they can embody specific forms of power and authority.

—Langdon Winner, "Do Artifacts Have Politics?" (1980)

During the early Apollo space missions, we were able to look back at Earth for the first time from a new perspective, viewing with our own eyes all of humanity and our environment. Previously, our observations were crude, limited to the boundaries of deserts and forests, the courses of rivers, the pockmarks of huge mines, or large-scale structures such as the Great Wall of China. In the decades since then, our shared powers of observation have grown exponentially. By turning the same tools we developed to satisfy our curiosity about the billions of stars beyond our solar system back upon the Earth, we have been able to explore and understand the profound impact of our influence on our planet. Many of the same technical achievements in digital imaging are now unlocking the microscopic wonders of life for all to share. Now nearly anyone can be equipped to make visible the "invisible" multitudes of microbes in the land and oceans that maintain Earth's livability to this day.

It is because of our shared tools and knowledge that we can say that July 2021 was the hottest month in recorded human history. These tools also enable us to report with high confidence that our oceans are acidifying, there is more carbon in our atmosphere than any time in the past three million years, the sixth great mass extinction of species in Earth's history is underway, and there is roughly four times more carbon captured in soils than above ground and fifty times more carbon in our oceans than in the atmosphere. These tools teach us that each human is an expression of an evolving genetic code intertwined with billions of microbes that cohabitate and coevolve with us. As we better understand our own microbiome, the more we can contemplate that we are individually a meta-organism within a meta-organism within a global feedback ecosystem built upon a foundation of carbon-based life.

We are now creating silicon- and carbon-based technology that is coevolving with us. Sixty years ago, we began to expand our powers of observation with satellites and other eyes-in-the-sky technology, giving us the ability to observe in "third-person view." We have now added increasingly complex senses to our external "nervous" system using silicon-based devices that translate waves into digital signals carried through a network of fiber optics, microprocessors, and sensors to make visible the complexity of life on Earth and our role in influencing it.

It is because of our capacity to measure these changes that many people have suggested that International Commission on Stratigraphy designate the current geological epoch as the Anthropocene—the era during which human activity became the dominant force on the planet. The idea originated with Paul Crutzen, a Dutch atmospheric chemist and Nobel Prize winner, who popularized the term in 2000 as a way of provoking a dialogue about the idea that humans have supplanted nature as the principal driver of change on Earth. In contrast, the epoch's predecessor, the 12,000-year-long interglacial period called the Holocene, was characterized by stable sea levels, a favorable climate, and low levels of human impact, at least from a geological perspective. Not coincidentally, this was the same period in which human society transformed itself from hunter-gatherers and land stewards to the more recent rise and dominance of the industrialized world on the landscape. As a starting point for the Anthropocene, a committee of geologists recently proposed July 16, 1945—the date of the Trinity nuclear detonation in New Mexico. Subsequent atomic blasts by various governments, at a rate of one every ten days for two decades, have created a global chemical signal that is detectable geologically.[1]

Whatever start date is ultimately selected, the physical evidence for the Anthropocene is widespread and easily recorded by our new tools, ranging from the accumulation of carbon dioxide in the atmosphere, rising rates of biodiversity loss, and the exploitation of natural resources to increasing deforestation, desertification, aquifer depletion, and wetland loss. We know that microplastics can be found in the Antarctic snow and in the deepest reaches of the ocean.[2] The measurable consequences of our activity can also be found in our cities and towns, landscapes that have been built in response to our evolving technologies (cities are optimized for automobiles rather than people, for example). Taken together, the abundant evidence of our impact on the planet is now verifiable.

Not surprisingly, Crutzen's concept caught on and has circulated widely in the scientific, academic, business, and nonprofit communities. A 2011 cover story in the *Economist* about the Anthropocene made headlines with its dire warnings about the implications of human-induced environmental crises. It was a public endorsement of the concept that soon spread across media, spawning books, websites, journals, and documentaries. In the summer of 2019, a report on agriculture and land use by the UN Intergovernmental Panel on Climate Change (IPCC) identified humans as the primary force behind land degradation, melting permafrost, higher rates of pest infestation, wildfire, and crop failures. Without urgent and transformational action directed at our food production system and land management activities, the authors said, all these trends will continue in coming decades.[3]

Call it the Bad Anthropocene.

Humans as Beneficial Organisms

There's another way to look at this new epoch, however, one that is hopeful and action-oriented and turns climate change arguments on their heads. If the rapid change we have observed in our lifetimes has been largely created by human action, then we have *already* proven that we can collectively change the planet's climate and function. This knowledge is empowering. It bolsters the argument that if we can change the climate in one direction, then we can change it back. While it is true that nearly anyone can burn down a barn, it takes a community to build one. We collectively put damaging chlorofluorocarbons (CFCs) into the atmosphere, detected the effect on the ozone layer, and then reversed course. The same can be said of lead in gasoline and other pollutants in the environment. And in the case of human health, we have shifted course on smoking in just a few decades. In each case, progress followed, though not immediately, when scientific analyses were coupled with societal approaches to action. There is always resistance to change, but we have learned that when systems are deeply understood, information and feedback can lead to action in a number of positive directions. When I first spoke at the Quivira Coalition Conference in 2011, the concept that agriculture had a role to play in mitigating climate change was novel, and yet just eight years later I saw the former president of the American Farm Bureau, Richard Stallman, introduce former Vice President Al Gore at a Foundation for Food and Agriculture Research event in Washington, DC, for a talk on the potential for regenerative agriculture to be a natural climate solution.

The editors of the science journal *Nature* put it well in 2011 when they wrote that the idea of the Anthropocene encourages "a mindset that will be important not only to fully understand the transformation now occurring but to take action to control it. . . . Humans may yet ensure that these early years of the Anthropocene are a geological glitch and not just a prelude to a far more severe disruption."[4] Echoing this sentiment, Crutzen wrote an article during the same year for the online magazine *Yale Environmental 360* in which he said recognizing this new period means we can shine a spotlight on our intellect, creativity, and the opportunities each offers for shaping the future. "We need innovations tailored to the needs of the poorest," he wrote, "for example new plant varieties that can withstand climate change and robust iPads packed with practical agricultural advice and market information for small-scale farmers. Global agriculture must become high-tech and organic at the same time, allowing farms to benefit from the health of natural habitats."[5]

Over the past decade, a wide variety of sophisticated, low-cost, biology-based regenerative solutions have indeed emerged—solutions that work with natural processes to restore and maintain healthy landscapes, improve habitat, increase biodiversity, capture atmospheric carbon in soils, improve water quality, mitigate floods and droughts, and improve the resilience of agricultural land and communities. These regenerative solutions are knowledge-intensive and culturally transformative. They link local and Indigenous experience with cutting-edge observational, analytic, and communications technology, resulting in improved ecosystem function that not only creates resilient food and fiber systems but also provides a host of other services that benefit life on Earth. The potential for using deep knowledge applied at scale, to act individually and globally, points away from the current Bad Anthropocene pathway toward a direction of continual improvement of ecological and social health—something we might call a *Good* Anthropocene.

Danger and Opportunity

Informed action is possible because technology is becoming less expensive. It is also nearly invisible, as miniaturization and ubiquity become a feedback loop in advanced manufacturing, creating the potential for tools to assist us in observing and shaping the environment in order to heal damaged land and feed people. However, the onrushing technological capacity to manufacture nearly anything anywhere and to communicate with everyone everywhere has not yet enabled land stewards to produce abundance regeneratively everywhere. One reason is philosophical: To succeed we have to know what to ask from our technology, and we must have the imagination and desire to put our collective capacity to good use. Technology can't happen in a social vacuum. Advances in artificial intelligence, for example, will accelerate our ability to interpret complex systems but not necessarily change the cultural and economic trends that led to the Bad Anthropocene. In fact, AI might make things worse. The same tools that are being used to interpret satellite images to identify the lands that might be most rapidly regenerated or to monitor fires in the Amazon are simultaneously being used to prospect for new mining operations and to target artillery. If we do not match advances in technology with social progress and innovation in governance, we run a big risk of creating a world organized by computers designed by and for a few powerful interests.

As Marc Andreessen, the billionaire venture capitalist who cofounded Netscape, famously said, "The spread of computers and the Internet will put jobs in two categories: people who tell computers what to do, and people who are told by computers what to do."[6] We are at a crossroads—if we do not consciously envision a better world and take steps toward it, then we will inherit a world created by algorithms that will codify a future that will not be recognizable or hospitable for us.

Joseph de Maistre, an eighteenth-century French lawyer and monarchist, put it bluntly when he warned in the wake of the French Revolution that "Every country has the governance it deserves." A charitable interpretation of Maistre's statement is that governance is a type of collaborative technology—things we chose to do together—and that positive and negative paths forward are possible, but improvement requires participation, inclusion, and innovation. Without that, we are not at the helm. Fortunately, the next phase of the Anthropocene is not etched in stone. We can move in multiple different directions depending on our goals. As Kae Tempest says in her song "Parables," "We've got the tiller in hand." It is our choice how we navigate into the future to "build bridges over this splintered land before the hourglass cracks and spills its sand."[7]

On our side are millions of years of biodiversity that evolved to find a way to grow on every acre of the Earth. Evolution is an incredibly powerful force with which we can align our efforts. Instead of destroying biological diversity, we can harness our technical innovation and creativity to regenerate natural processes and produce healthy food while restoring landscapes and thereby creating a positive legacy. However, we will only succeed in countering the current extractive trajectory if knowledge about how to restore and employ natural systems for the benefit of all life is democratized. Knowledge needs to be made available in such a way that any person on the planet has access to, and can participate in the creation of, the best local information to regenerate the land under our control. By sharing knowledge and working collaboratively we can work with natural systems and substantially raise our chances for restoring our well-being and that of our fellow travelers on Earth.

There is a plausible danger, however, that if the Good Anthropocene concept is not developed within the regenerative framework, it will become another example of the same industrial mindset that created so many large-scale unintended consequences in the first place. For example, the climate crisis has already encouraged proponents of planetary geoengineering, as well as advocates of techno-utopian worldviews, to make a case for radical, untested, and risky industrially based strategies to "cool the planet." This is not what Paul Crutzen had in mind. "To the dismay of those who first proposed it, the Anthropocene is being reframed as an event to be celebrated rather than lamented and feared," wrote ethicist Clive Hamilton. "Instead of final proof of the damage done by techno-industrial hubris, the 'ecomodernists' welcome the new epoch as a sign of man's ability to transform and control nature."[8] As a consequence, he argues, an important opportunity to confront global capitalism's role in our predicament, as well as humankind's myopia, is being ignored. In other words, embracing the Anthropocene in this way won't change our behavior or acknowledge planetary and biological systems that underpin life on Earth.

Hamilton has a point. The framing of a Good Anthropocene in which humans can adapt and prosper in an increasingly warmer world is a dangerous fallacy unless we simultaneously create governance and market structures built on informed values and improved environmental outcomes we can measure. We will also need a keen and skeptical sense of nation-states' roles in attempted social engineering through multinational organizations such as the World Bank and the Food and Agriculture Organization (FAO). In later chapters, we will explore how, because of ubiquitous communications, we now have far more ways of organizing and governing large-scale efforts than through traditional nation-states or corporate structures. Moving forward, we will need a humble but realistic sense of our substantial capacity to affect our commonwealth of nature both positively and negatively. In the past, unintended environmental and social costs—called *externalities* by economists—were discounted or simply ignored. They cannot be ignored any longer. Fortunately, an economic theory that aspires to absorb externalities and create efficient markets driven by the latest information, accessible to all, is no longer an impractical fantasy. It now squarely rests in the realm of the possible in the public imagination.

We have already learned by hard example that interfering in complex systems can be disastrous on local and global scales. However, given the huge environmental and climate challenges pressing down upon us, do we have a choice? No. Besides, as Crutzen suggested, the idea of the Anthropocene need not subsume itself to business-as-usual, extractive, industrial, and capitalist economic thinking that lacks feedback and full financial accountability. The natural world is the largest dynamic wealth-generating force on Earth, transforming sunlight into life on a scale and level of complexity that dwarfs industrial production. Living systems are both negatively *and* positively influenced by our management, which means regenerative agriculture can create a new future, just as nature created a livable biosphere and made our very existence possible. As a curious and ambitious species, we have the power to not just extract scarce resources but rather to use our knowledge to restore and regenerate natural systems. Backed by powerful analytics and high-resolution data, we can restore health to the environment and improve our prospects while shifting from an economy based on scarcity and competition toward one based on abundance and nonrival common benefits.

The new narrative we need not only sees humans as positive agents for change but also lays out a road map with clear goals and measurable mileposts for achievement—our efforts are linked together, information is shared, and our goals benefit all life on Earth. In this new narrative, we pull back from the brink of disaster, no longer ignorant of our actions or their consequences, and work toward restoration, resilience, and abundance.

Let's call this new narrative the Great Regeneration.

Actors on a Stage

On October 4, 1957, the Soviet Union launched a shiny, silver satellite the size of a beach ball into space and changed the world forever. This first artificial satellite in history, Sputnik 1, circled Earth in a low orbit for nearly three months, passing over many population centers, before succumbing to gravity and perishing in the upper atmosphere. For the first few weeks of its short life, Sputnik issued radio signals from its slender antennae that were easily picked up by listeners on the ground, including amateur radio enthusiasts, beginning an ongoing transformation of how we see ourselves in relationship to one another and to our planet. This satellite triggered the Space Race, culminating in astronaut Neil Armstrong's famous first steps on the moon, and in so doing provoked a sea change in technology as well as a new understanding of the Earth. The economic and scientific advances ushered in by satellite technology quickly became global, for better and for worse. This is one reason why social critic Marshall McLuhan's comment, made in 1974, was so prescient: It was an observation made just a little more than a decade after Sputnik ushered in an era where no human on the planet was a bystander any longer. Now we are all actors in a grand drama involving many intersecting parts: rapid globalization, rising standards of living, longer life spans, the proliferation of increasingly sophisticated technology, the impact of larger and more urbanized populations, ever larger scales of natural resource destruction, and the existential threat of climate change.

There is another way McLuhan's comment rings true. As the first artificial satellite, Sputnik was the harbinger of all the space-based observation and communication technologies that followed, many of which have profoundly changed not just how we see the world but also who is able to see us. By 2021, more than 11,000 satellites had been launched into space with over 7,000 still in orbit.[9] The trend now is toward a vast quantity of less-expensive satellites, launched with reusable rockets that are no longer the exclusive domain of state superpowers. As McLuhan's comment implies, there is literally no place to hide anymore. The planetary stage on which we live, work, and sleep, is now subject to scrutiny from a wide variety of highly accurate observation technologies, evidenced in spring of 2022 by the disturbing satellite images of dead bodies on the streets of Bucha in Ukraine, victims of Russia's brutal assault.

Nearly everything on Earth's surface is now visible, measurable, and sharable. I recently took a Google street view tour of Buenos Aires and Hong Kong and could point out to my family the restaurants and cafés that I had frequented twenty years earlier. Exploration is no longer the discovery of unknown lands, but the development and sharing of human understanding of the vast biodiversity and complex biogeochemistry of the planet, from the billions of algae, fungi, and bacteria within complex ecosystems to the watershed, landscape, and biosphere levels. In this sense, all living things are now actors on the stage.

What we choose to do with this deep knowledge, as well as the rapidly evolving technology that helped to create it, is a profound question. As a species we have acquired the powers of both the superheroes and supervillains of popular culture. While the villainous aspects of advancing technology, such mass surveillance, are increasingly prevalent in society, the heroic aspect is often overlooked or underdeveloped. For example, the launch of microcontrollers and digital sensors into space in recent years has started the process of creating a silicon-based nervous system that can observe and document the already highly networked carbon-based life system of Earth. Increases in resolution and sensory capacity mean we are able to see in more wavelengths and at higher detail and are beginning to witness what was previously invisible.

As an analogy, consider the unconscious functions of the human body. We process sugars, breathe, and physically and chemically balance ourselves without thinking about it. However, if any of these systems stops functioning, even for a short time, we know instantly. Let's call these *utilities*. Our body's neurons are not particularly useful to us without the ability to interpret the signals to make meaning and put our conscious interpretation into purposeful action with our muscles. Let's call these intentional and informed *actions*. It's the same with silicon signals from satellites. We have the equivalent of a vast network of sensory system functions that are connected to analytic systems that can translate wavelengths into digital outputs (aka utilities). Together these systems can make meaning of the signals and share them so we can transform information into knowledge that can then be put toward collective action. In just the past few years, these large-scale systems, built on an ecosystem of utility services, have become a reality at the landscape scale and have enabled us to tell data-driven stories that explain things such as the boundaries of Gabe Brown's farm or the recovery of my own farm from drought and flooding.

This story begins with the pioneering Landsat remote-sensing satellite program, an ongoing, uninterrupted survey of the Earth's surface from space that has delivered visually dramatic and highly detailed imagery, literally redefining the way we look at our planet. The idea for a comprehensive record of Earth from space was proposed in 1966 by the head of the US Geological Survey, William Pecora, who was inspired by images of the planet being generated by the Gemini space program. Until then, remote surveillance had been conducted from high-altitude planes, including many secret missions by the US Department of Defense. The prospect of a civilian-led survey of the Earth's surface for public and scientific purposes met political opposition in Washington, DC, as well as objections from foreign governments. When US Secretary of the Interior Stewart Udall threatened to undertake the project anyway, NASA jumped in with its own plans, culminating in the launch on July 23, 1972, of Landsat 1 (originally called Earth Resources Technology Satellite or ERTS-1).

The goal of the program was to map global terrestrial systems at medium- to-high resolutions with repeated multi-spectral measurements, thereby creating a shared database for scientific, land-planning, and commercial use available to any citizen. What it achieved was nothing less than a revolution in our understanding of the Earth, creating the foundation for document-ing the Anthropocene as it unfolded. For example, its spectral analysis of changes in vegetative cover, such as rates of evaporation and photosyn-thetic absorption, in combination with land-use patterns, transformed our knowledge about how ecosystems function at large scales under the impact of human activities. This new type of observational data was one reason why NASA Administrator James Fletcher predicted in 1975, as Landsat 2 was being prepared for launch, that if one space-age development would save the world, it would be Landsat.[10]

The use of satellite data to analyze natural resources and land use expanded in subsequent years to include farming, forestry, pollution, and water management, among many other subjects. As the database grew, year-to-year changes in land cover became more detailed, especially as eye-in-the-sky technologies grew more sophisticated with each new satellite. Landsat 3, launched in 1978, had cost nearly $1 billion in today's dollars and had a 75m resolution (images were readable to an area of 75 square meters), which enabled the identification of large field boundaries. Landsat 8, launched in 2013, is capable of 15m resolution, and Landsat 9, launched in 2021, provides even greater precision. Meanwhile, by 2003 the cost of image chips had dropped by 99 percent to about one cent per sensor pixel. By 2019, software enabled unmanned aerial vehicles (UAVs) to produce imagery with 5cm resolution for a few hundred dollars. The refinement of the imagery resolution means we are able to do more than just pick out individual field boundary edges; we can now identify individual plants, flowers, leaf shapes, and colors.

These developments were part of a larger trend taking place in society. Independent of external environmental costs to produce new digital storage, imaging, and processing, more power is generally being squeezed out of less material. There has been a convergence of lower production and purchasing costs, resulting over the course of just a few decades in the emergence of ubiquitous communications technology and the rapid conversion from analog to digital in multiple sectors, followed by the vast expansion of storage capacity and processing power, which has led to a nascent "Internet of Things." These phenomena are defined as the extension of the internet to physical devices and everyday objects that can be monitored and controlled remotely. The possibility of billions of interconnected devices around the world communicating together, once science fiction, has become a reality in a fraction of our lifetime.

The Landsat program played another important role in demonstrating the usefulness of data and knowledge to modern society, though this one happened by trial and error. In 1984, Congress ordered the Landsat program to be privatized as part of an ideologically driven push by conservatives to transfer government services to industry. The result was a steep increase in the cost of Landsat's data to the public—from roughly $650 to $4,400 per image—pricing out many researchers and other citizens. There were other unexpected issues with the program, as well. This experiment in privatization came to a head in October 1993 when Landsat 6 failed to achieve orbit due to a malfunctioning fuel line, threatening to interrupt the twenty-year run of continuous data collection. Fortunately, Landsat 5 continued to function. Under pressure, Congress returned control of the program to NASA, which successfully launched Landsat 7 in 1999. Image costs quickly fell to previous levels. Then in 2008, NASA decided to make all Landsat data free to the public.[11] Demand rose from less than a million downloads a year to 100 million in 2020.

Fast forward to today. Detailed, easily accessible imagery of the Earth is ubiquitous, as exemplified by Google Earth, a highly popular eye-in-the-sky service available on the internet which allows users to browse Earth imagery for free. As observation and communication satellites have multiplied overhead, an equally revolutionary force has been taking place on the ground, mirroring the technical ecosystem in the sky. Much of the same digital imaging and sensor technology developed for use in satellites—such as digital cameras, spectrometers/colorimeters, accelerometers, and compasses—have also dropped in cost to the point where smartphones and tablets, which have vastly more computing power than the first vehicles launched into space, are now being used to map our lives and environments in real time, sometimes without our consent.

The full potential for observing and documenting life on Earth by connecting the on-the-ground data collection with remote observations is coming into focus with the increase in computational power and the emergence of machine learning, both of which are required to interpret the vast complexity of our shared world. In this way, carbon (life) and silicon (data collection and communication) are being combined to provide an unprecedented opportunity to direct our capacity to nonauthoritarian ends and to understand and thoughtfully manage our world, from the microscopic to the global. As Langdon Winner, a noted political theorist, said, these objects we create, this technology, is a product of the values and resources we place into them. We can make soil sensors and biodiversity monitors more ubiquitous than facial recognition, robotic guard dogs, or kamikaze drones. The technology we produce is not neutral or inevitable. We chose what tools we produce and use. It is a choice that reflects our values and the world we want to live in.

From the Invisible to the Visible

One of the most important discoveries revealed by Landsat data and other scientific aerial surveys is the dramatic human impact on the Earth's natural systems, which began to escalate around the time of Sputnik's launch into space. After World War II, the economies of industrialized nations quickly expanded, requiring additional raw materials, such as coal, oil, and lumber, on scales that accelerated ecosystem devastation, pollution, and overharvesting. By the time Sputnik completed its first orbit, many destructive trends were starting to grow at ever-increasing rates, signaling a fundamental shift in the relationship between humans and the planet. Within the short period during which measurements have been made, the effects of this shift have become profound.

"It is difficult to overestimate the scale and speed of change," wrote Will Steffen, the principal author of a report on the "Great Acceleration," from the International Geosphere-Biosphere Programme (IGBP) and the Stockholm Resilience Centre. The researchers discovered that within a single human lifetime, humanity became a planetary-scale geological force. "When

Socio-Economic Trends

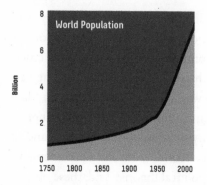

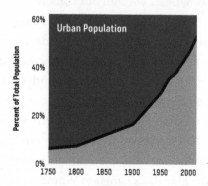

Earth System Trends

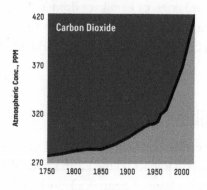

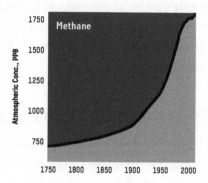

Technology Trends

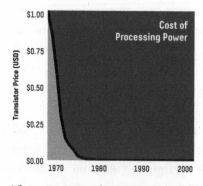

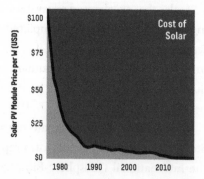

These Great Acceleration graphs highlight human and Earth system trends from 1750 to 2010. *Creative Commons.*

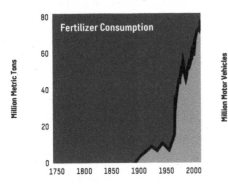

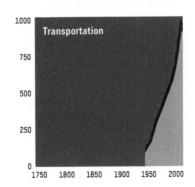

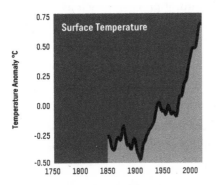

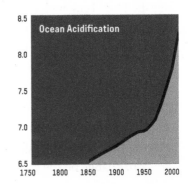

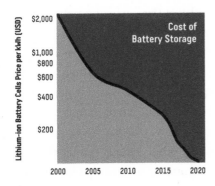

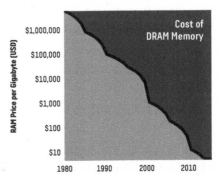

we first aggregated these datasets," Steffen wrote, "we expected to see major changes, but what surprised us was the timing. Almost all graphs show the same pattern. The most dramatic shifts have occurred since 1950. We can say that around 1950 was the start of the Great Acceleration. This is a new phenomenon and indicates that humanity has a new responsibility at a global level for the planet."[12]

"Of all the socio-economic trends only construction of large new dams seems to show any sign of . . . slowing," added coauthor Lisa Deutsch. "Only one earth system trend indicates a curve that may be the result of intentional human intervention—the success story of ozone depletion. The leveling off of marine fisheries capture since the 1980s is unfortunately not due to marine stewardship, but to overfishing."

Humans have become the dominant driving force for change, even altering the basic functions of important natural processes such as the cycling of carbon beyond the normal range of variability. Documented by global monitoring of environmental systems, indicators illustrate how widespread the Great Acceleration has become. These include trends that approach or exceed in magnitude some of the great forces of nature. The extent and rate of alterations to the natural functioning of planetary physical, biological, and chemical systems are unprecedented in human history and our impact, especially as a consequence of economic activities, has become the primary driver of change on the planet.

This rapid rise in the rate of use and abuse could also be called the Great Extraction. It is not a coincidence that many dystopian visions in science fiction focus on mining communities or colonies. Even the *Star Trek* series explicitly uses the language of "the final frontier," which communicates a concept of civilization based on scarcity that is constantly on the move to find new resources in ever-expanding circles of colonies. Mining permanently removes a resource without replenishing it, resulting in a constant search for the next ore load to exploit once the original source has been depleted. Industrial agriculture, with its focus on mineral inputs of nitrogen, potassium, and phosphorus, is also mining-based. Copper, coal, oil, diamonds, gold, silver, guano, natural gas, fossil water, and many other natural resources have all been extracted and depleted over the centuries, a pattern that has expanded recently with the development of new exploration and extraction technologies. Previously hard-to-reach resources are now being mined at prodigious rates, and the same new technology used to map environmental destruction is often employed to map and find resources more efficiently. Fracking is one example. The end result is the same: a cycle of eventual exhaustion of the resource and a transition to a substitute resource. It is worth pointing out that sophisticated technology is not required to create large-scale destruction. Throughout history, civilization after civilization has collapsed as a result of agricultural and environmental destruction created by "low technology" such as the plow, which caused widespread soil erosion through overuse.[13]

When exploitative ideology is applied to biological resources, a similarly predictable result follows. The poaching of elephants for their ivory has decimated elephant populations. Whales, cod, rhinos, and old-growth forests have experienced similar trajectories. Add climate change and the rising global population of humans, both of which are having devastating effects on habitats, and you get a prescription for a biological crisis. Technology, such as drone imagery, used to monitor endangered species is now inexpensive enough to be used by poachers, as well.

In May 2019, a landmark report from the Intergovernmental Science-Policy Platform on Biodiversity and Ecosystem Services (IPBES), compiled by 145 experts from 50 nations and based on 15,000 scientific and governmental sources, concluded that one million animal and plant species are now threatened with extinction, and that biodiversity is declining around the world at rates unprecedented in human history. "The health of ecosystems on which we and all other species depend is deteriorating more rapidly than ever," said Sir Robert Watson, the chairman for the group. "We are eroding the very foundations of our economies, livelihoods, food security, health, and quality of life worldwide."[14] It is worth highlighting some of the report's points about agricultural use:

- More than a third of the world's land surface and nearly 75 percent of freshwater resources are now devoted to crop or livestock production.
- The value of crop production has increased by about 300 percent since 1970; raw timber harvest rose by 45 percent; and sixty billion tons of renewable and nonrenewable resources are now extracted globally every year, nearly double since 1980.
- Soil degradation has reduced the productivity of 23 percent of the global land surface; global crops are at significant risk from pollinator

loss; and 100 to 300 million people are at increased risk of floods and hurricanes because of the destruction of wetlands and other mitigating natural features.

- In 2015, 33 percent of marine fish stocks were being harvested at unsustainable levels with just 7 percent harvested at levels lower than what can be sustainably fished.

These concerns are not new. The roots of the conservation movement that arose in the nineteenth century came out of reactions to colonial excesses. In North America, hunting clubs began to worry about declining game populations. Soon, a general alarm about wildlife sounded as human impacts became more profound, symbolized by the extinction of the passenger pigeon, whose population once numbered in the billions. The last wild bird was shot in 1901. Two years later, President Teddy Roosevelt created the nation's first National Wildlife Refuge by executive order, even as the formation of national parks excluded Indigenous people from their own land. However, conservation strategies developed in the first half of the twentieth century, including national parks, forest reserves, and wildlife refuges, are inadequate in stemming the rising rate of extinction represented by the Great Acceleration, which has continued despite conservation programs.

Perhaps that is because the movement failed to understand conservation beyond preserving "wilderness" and slowing environmental degradation, rather than creating a broader vision that emphasizes support of these functioning natural systems and acknowledges the people who stewarded those systems for thousands of years. The fundamental shortcoming of this mindset is to believe that nature is separate from humans rather than to consider our direct participation and potential as *part* of nature. The challenge is not to carve out a corner of nature within an extractive economy, it is to fundamentally reform the economy to value the regenerative and expansive aspects of nature as the foundation for industry, arts, and commerce. According to Sir Robert Watson, the solution to this expanding crisis can be nothing less than "transformative change . . . a fundamental, system-wide reorganization across technological, economic, and social factors, including paradigms, goals and values."[15]

The documentation of habitat and wildlife above ground is critical to the Great Regeneration, but understanding the unseen microbial universe in the soil is also key to assessing our successes or failures. Our ability to document the Great Acceleration provides proof that we can use technology to place ourselves in nature and not break away from it. The technology we use to visualize the scope of the problem also provides a guide for how to reverse the damage. While intuitive, progressive, and ethical action at the individual farm or landscape level might seem to make all this detailed documentation unnecessary, the simple reality is that science, microbiology, and biogeochemistry are almost impossible to implement by "feel" or instinct. Great advances in science are often *not* intuitive and become evident only when observations are structured and combined to find the unexpected. Effective management at scales needed to reverse the Great Acceleration can only be derived—and verified—by one thing: *greater shared knowledge*. This knowledge, which goes beyond the capacity of individuals to hold alone, is enabled by shared institutions and can influence culture and action at a landscape scale.

The power of observational tools lies in their neutrality. The data they collect allows us to collectively create new values, or reinforce old ones, based on trusted knowledge about our impacts upon our environment. The classic examples of natural resource market failures such as overfishing or overgrazing are often referred to as a tragedy of the commons. However, the script can be flipped. There are numerous historic examples of Indigenous governance systems that successfully managed common resources over generations. And now that we have clear data, it might not be surprising that what we implement will have echoes of Indigenous governance systems for stewarding the underlying services performed by healthy rivers, grasslands, successional forests, estuaries, and fisheries.

Climate change is perhaps our largest opportunity to turn a tragedy into a triumph of the commons. The balance of carbon in soil, plants, sea, and atmosphere serves as a global indicator of our local efforts to shift toward more life rather than less. There is vastly more carbon in soil and plants than in the atmosphere, so seemingly small changes can have dramatic effects. Efforts to increase biological diversity and complexity offer a viable and proven approach that can also provide ecological resilience, restoration, and regeneration—a reversal of the Great Extraction. Life has refined an adaptive approach, proven over many millions of years of evolution, that is represented in the biodiversity adapted to nearly every location on Earth, which in turn has contributed to an oxygen-rich climate that is advantageous to life on Earth. Climate change's multiple, deleterious, and rapidly expanding threats to our well-being and the natural world are the essence of the Bad Anthropocene. However, approaching this crisis from the perspective of the Great Regeneration—as a shared human challenge aligned with powerful natural forces—means we can meet this challenge with informed intent and hope.

CHAPTER 2

Our Commonwealth of Knowledge

The goal of an encyclopedia is to assemble all the knowledge scattered on the surface of the Earth, to demonstrate the general system to the people with whom we live, & to transmit it to the people who will come after us, so that the works of centuries past is not useless to the centuries which follow, that our descendants, by becoming more learned, may become more virtuous & happier, & that we do not die without having merited being part of the human race.

—DENIS DIDEROT, *Encyclopédie* (1751)

When the morning sun emerges on the horizon we can see a few minutes into the future. Physics provides an elegant explanation—the atmosphere refracts light over the horizon in such a way that we see the sun shortly before it is actually in our line of sight. The physics is the same whether we're standing in a field in Nebraska or Malawi, but it becomes clearer if we head over to the local airfield and fly up to 10,000 feet. There, the horizon exposes the long curvature of the Earth stretching out ahead, allowing us to experience the same effect from a higher perspective. Even as the expanse of our view extends, we can also see that more is hidden just over the horizon. We gain context from a new point of view. If we were to ascend to 90,000 feet, above the operating altitude of most private aircraft, the shape of the Earth and the atmosphere both become visible, the latter as a thin layer of gasses blanketing our life-sustaining biosphere, held in balance by life on Earth. We can see now that a fragile boundary between space and

Earth gives us our distinct blue sky, which is made possible by our planet's unique position in the solar system. While looking out across the arch of the receding horizon, we can descend again and pivot to look out and down from the side window of the plane, down to the fields 10,000 feet below. The details of the plants visible at ground level blur into patterns of color that can be read almost like pixels in a larger palette of greens and soil tones that create even larger patterns in a shifting mosaic of life overlaid on a landscape of ridges, valleys, and rivers.

The vision isn't static. The full story of the terrestrial surface of the Earth involves the breathing of the land into the biosphere, which reflects and refracts bandwidths of visible color to our eyes. To better understand this story, we need to fly back down to inspect the details that make up the patterns within the forests, rivers, ridges, and fields. We must travel deeper into the fields and zoom down below the surface of the ground and into the soils and plant roots, where hidden metropolises of bacteria and fungi thrive. It is in these living, breathing spaces between broken rock, sand, silt, and clay that fungal communication networks transport messages and nutrients underground among plants across vast distances. If we take a handful of rich soil dug in a Malawi garden, for example, and place it under a simple digital microscope linked to the screen on our phone, we can observe the billions upon billions of microbes that tell a story that takes us back in time to the origins of life on Earth. When we take a sample of those microbes and analyze them in the field with a stapler-sized USB-connected DNA sequencer, we can look even deeper using a near-instant genetic analysis of this living microbiome. The DNA structures of these most basic microbes invite us to explore the shared architecture of all life forms and to contemplate the evolutionary processes that propel life toward ever-increasing diversity and complexity.

I have had the privilege of using these tools in fields on my own farm. You may be wondering what astrophysics and genomics have to do with agriculture? Isn't agriculture *just* a business, something that farmers and ranchers do rather mechanically to produce food, fiber, and fuel from their land? Farming isn't rocket science, right? No, it's *more* complicated! I propose that our understanding of agriculture must include and expand beyond the highly complex history of our interaction with microbes and our vast transformation of the biosphere through our actions to produce food, fiber, and fuel.

However, I would also suggest that agriculture and its outcomes have always been a quantifiable measure of our success or failure as a species to use our collective knowledge of nature to overcome scarcity by harnessing the life around us to create abundance where it wasn't before. Agriculture is sociology, technology, economics, physics, chemistry, biology, and history—all rolled into one.

By taking a longer view, we can see the arc of the horizon of our agricultural past, which gives us the ability to put our endeavors in context and to see what's possible for the future just over the horizon. By using new tools of scientific collaboration, from genomics to microscopes, satellites, and radio telescopes, we can create order within these levels of perspective. We can combine what we see with our own eyes at ground level with the pixilated patterns produced by satellites and microscopes, resolving images into sharper focus as we move closer or farther away—looking both forward and backward in time. Our tools, once in the realm of science fiction, are now equivalent to an extraordinarily powerful zoom lens accessible to all of us—and we are just beginning to learn where and how to focus our lens and what to do with the knowledge our powerful new observational capacities provide. In just the last decades we have begun to explore the boundaries of these new powers to learn if we can indeed take collective action to stop planetary degradation, counter scarcity, and instead use these tools to help us move toward regeneration and abundance.

The Knowledge Commons

As we survey our observed landscape, from the microbiome to the biosphere, and consider how we choose to manage it, we are tapping into our shared knowledge commons. A commons is defined by the cofounder of the Public Knowledge group, David Bollier, as "The wealth that we inherit or create together and must pass on, undiminished or enhanced, to our children. Our collective wealth includes the gifts of nature, civic infrastructure, cultural works and traditions, and knowledge."[1] A knowledge commons is how communities build upon each other's work and how we express the social technology of collaboration as a species. These commons collectively form our commonwealth of knowledge.

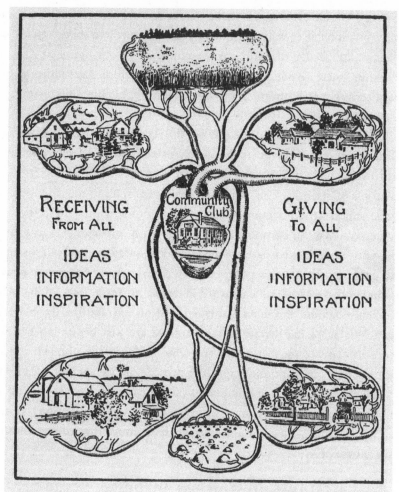

FIG. 281. The community club, rightly organized and conducted, can serve as the heart of the community through and by which a stream of inspiration and encouragement is sent to every member.

The *Encyclopedia of Practical Farm Knowledge*, published and widely distributed by Sears and Roebuck 1918, illustrated the agrarian club as central to a circulatory system of knowledge, information, and inspiration.

Collaboration has proven crucial to the creation of civilizations and the well-being of *Homo sapiens* over millennia. The expansion of collaboration technologies, from simple illustrations and diagrams in books to sophisticated digital software, enables sharing an emerging understanding of how the world actually works. A knowledge commons translates how the world works from the infinite language of things-as-they-are to a symbolic understanding that is comprehensible for humans. We can effortlessly share digital images of family gatherings. Radio waves send our voices everywhere. None of this is based on intuition but rather on cumulative insights that are shared in ways such that others may build on our work and contribute to the commons. A commons can exist, like gravity, regardless of our awareness of its existence or our ability to explain exactly how it functions.

According to Bollier, a commons is a self-organized system by which a community manages a resource in a collective manner that ensures its regenerative, long-term viability while preserving shared values and local identity. There is little or no reliance on nation-states or outside interests. Control is strictly in the hands of the community. Along with their historical rights, a commons often has legal authority to operate as a collective. Far from being an anomaly, commons structures stretch back thousands of years as the dominant social tool for managing hunting and foraging grounds. Commons structures span most of human history, and they take many forms: Livestock grazing associations in Mexico, called *ejidos*, are composed of multiple families within a geographically defined area. In northern New Mexico, there is a 400-year tradition of sharing water from a communal ditch, called an *acequia*, for crop irrigation. Similar systems are found in Nepal. In the United States, the Freedom Farm Cooperative, founded by civil rights activist Fannie Lou Hamer, drew upon a history of the Black commons and ag groups. These combined efforts formed the foundation of many of our contemporary cooperative agricultural structures.

Numerous other resource-based commons, both of modern and ancient origin, are still used around the world to manage grazing, forests, water, foraging, and hunting and fishing habitats. Their goal, as Bollier points out, is to pass collective wealth, knowledge, culture, and land as intact as possible to the next generation, often increasing its value—and not necessarily in monetary terms. This is accomplished by protecting collective access to the resource, a decision-making process that puts the interest of the community first, and blocking outside forces that could degrade, remove, or privatize the resource or cultural tradition. The so-called "tragedy of the commons" happens when one or more of these safeguards breaks down or is willfully ignored by those who are motivated by extraction, resulting in damage to the resource's value.

There are commons that belong to every human and have intrinsic value to all life. The Earth's atmosphere, for example, is one of the most important shared resources on the planet, and its continued functioning and stability is essential to all life. The knowledge of shared resources and how they function is also a form of a commons to be carefully stewarded. Unfortunately, we have been treating critical environmental public commons as dumping grounds for excessive amounts of carbon dioxide and other greenhouse gasses like methane, partly because these commons are invisible to us individually. Unlike an important fishing ground defiled by a massive oil spill, much of the degradation to these public commons is far more difficult to detect or to act upon. Other collective resources that can be considered part of the global commons, both as a social structure and from a biological function, include the oceans, biodiversity (above and below ground), and genetic diversity.

To extend the concept further, we might also see human history and diversity as other sorts of cultural commons that go far beyond what might be designated by the United Nations as World Heritage Sites. However, as Bollier points out, there is no commons without the act of *commoning*—the practices we use to manage a resource and the values that guide them. The goal of commoning is to ensure collective benefit, whether for humanity or nature, and preferably for both. Take the metropolises of microbes mentioned earlier. A single handful of healthy soil contains billions of microbes. If we think of them as a commons and manage them for health and resilience—for regeneration, in other words—then the collective benefit for food production, plant diversity, and other cascading effects is huge and unending. The commoning we do as land stewards—and consumers, by extension— must be directed at these metropolises in order to make everything grow regeneratively. Bollier says a commons is characterized by bottom-up participation, and it doesn't get any more *bottom* than soil microbes.

Elinor Ostrom, a Nobel Prize–winning economist, studied real-world commons and their governing structures. She discovered many examples of prudent, cooperative behavior that enabled a commonly held resource to be managed sustainably for generations without a need for external government control or privatization. Her research on collective self-governance of open-access lands took her to Africa, where she studied grazing lands held in common; to Nepal, where she analyzed cooperative irrigation systems; and to fishing grounds off the coast of Maine. She discovered that successful, long-term management of a commons is the result of diverse interactions among cultural traditions, political arrangements, religious and social beliefs, as well as economic motivations. Not every commons she saw was perfectly managed, and some suffered from resource abuse, but none had been destroyed by short-term self-interest.

Ostrom determined that successful commons around the world exhibit similar principles, including:

- clearly defined goals
- local control
- effective communication among members
- trust
- reciprocity
- fairness
- a collaborative decision-making process
- mechanisms to resolve conflict
- accurate monitoring of the resource
- a graduated scale of sanctions for violators
- respect by authorities at higher levels for a community's various rights

In other words, over time a local community would work out internal rules for its governance and growth and enforce those rules on its own. It's not theoretical. Ostrom documented the successful implementation of these principles in case study after case study, proving that the rules worked in practice. The tragedy only occurred, she argued, when an outside group or external force—political or financial—intruded on the resource or interfered with its management for their own selfish interests. (The ongoing deforestation of the Amazon by non-Indigenous groups for short-term financial gain is a modern example.) Not all elements have to be local. Nonprofit organizations, technical experts, and others could contribute to the effective management of a commons, Ostrom noted, but they had to be respectful of local cultures and conditions and abide by community decision-making.

The Key: A Knowledge Utility

If our shared knowledge is to be translated into action—continually storing and distributing its information—it requires infrastructure, or what I will call a *knowledge utility*. The infrastructure is not simply the internet, though the fiber optics, switches, and servers of the internet do provide a key piece of the physical infrastructure. However, a knowledge utility is the specific parts of the physical and social infrastructure stewarded to deliver trusted, aggregated, and curated knowledge that can be universally accessed for our specific purposes. Akin to a modern-day electric utility, distributing power across a landscape, a knowledge utility provides shared information so that people can make better decisions and manage land for the common good, thereby unlocking the regenerative economy and supporting democratic values. An electric utility delivers power to illuminate every home and allows us to read at night when it otherwise would be dark out, while a knowledge utility facilitates the exchange of ideas, information, and inspiration that would otherwise be inaccessible to us.

There is a big difference, however: Power generation, much like an FM radio station, has historically been one-way to the consumer. In a knowledge utility, the communication is two-way, with an almost infinite number of connections flowing in both directions. Like an electric utility, the voltage and plugs must be standardized and accessible to people so they can develop tools that will work when they are connected to the outlet. Anyone who has traveled internationally knows the challenges of power cord adapters. Imagine the challenge if each city or neighborhood had its own proprietary electrical plug and home voltages ranging from 460V AC to 5V DC. Buying an appliance for your home would require a wide assortment of adapters, and charging a phone at a friend's house would be even more challenging than it is today.

A knowledge utility is a core innovation necessary to provide access to the knowledge commons in a useful form to everyone, everywhere. First attempts at this ambitious type of project—such as the effort to catalog and map global soils and land cover by organizations such as the Food and Agriculture Organization (FAO) and other UN agencies—relied on publishing papers as well as centralized governance and funding structures. These efforts were costly and stopped far short of being able to provide plant, soil, technological, and economic information to all land stewards globally.

The best modern example of a large-scale, decentralized collaboration in knowledge commoning is perhaps the open-source consortia and foundations such as Apache, Mozilla, and Linux, which create and curate much of the software that makes the internet function. Apps that we use every day are built on shared software libraries with thousands of contributors and are stored and shared on servers called repositories. Open-source development is a decentralized and publicly accessible software design process that can be modified and shared by multiple users. Software development kits enable developers to build on and repurpose code that has been shown to work well in new applications. Similar to sharing blueprints for a building, open source allows the source code to be inspected and enhanced by anyone who is able to understand how software is constructed. These decentralized processes enable a radically different approach to scale and governance than was possible in the 1950s vision of UN agencies aggregating all the world's agricultural data. The open-source model radically changes the potential for the next generation of agricultural knowledge utilities that are being envisioned by organizations such as the AgStack Foundation, which is a project of the Linux Foundation.

Another analogy is the code of life, DNA, which represents the Earth's original open-source project. Theoretically, genetic code is nearly universal—the same snippet of DNA, in most cases, will code for the same amino acid regardless of the organism or source. For example, in both humans and bacteria, a codon made of three thymine DNA-letters will code for an amino acid. We now know that all forms of life share the majority of our genetic code. An oak, a bat, a grasshopper, and a fungus all share the same code for cellular functions. This code, like a digital photograph, can be reproduced billions of times with almost perfect accuracy. It is shared because it has undergone the review process of natural or artificial selection and because it has been shown to work. Nature shares code through reproduction, and innovation and adaptation happen at the edges. For humans, our biological evolution has been a blink of the eye in comparison to many other species, and we are just now beginning to mature our ability to share and build knowledge that can bridge across cultures, geographies, and languages.

Open source relies on peer review and collaborative production to sub-stitute for natural selection and creates rapid lifecycles through continuous releases. This process encourages innovation and flexibility and creates longer-lasting designs as a result of their development by a community instead of a sole author or business. More involvement by more people means increased security and stability. It is not an accident that much of the mission-critical software that runs the internet is open source. An open approach forms the foundation of communities centered on the software design and project goals, and the trend extends beyond software. In 2014, Tesla open-sourced its patents and in just a few years became worth as much as the combined market cap of the nine largest car companies globally. More important, open source is a *philosophy* and a structured practice of communi-cations and documentation focused on the democratization of information and collaborative problem-solving.

Understanding the potential to apply open source more broadly requires a bit of background. The concept of open source echoes the work of historical encyclopedias and their modern wiki equivalents. In terms of the rate of growth, the number of publicly accessible reposi-tories and code libraries has transitioned quickly from a few thousand projects to millions. Open source is now the predominant method of software development, even if the general public is unaware of the shift. A quaint idea easily dismissed in the early 2000s when Linux was initially going up against the Microsoft Windows operating system monopoly, just twenty years later, following the maturing of the internet economy, open-source development is now ubiquitous and used to create a wide range of goods and services, including spacecraft, automotive operating systems, video editing, photo storage, and mobile apps. Android-based mobile phones—based on the open-source Linux operating system—make up the vast majority of smartphones on the planet. Even Microsoft, once the emperor of proprietary code, has embraced Linux for cloud services, which make up an increasing portion of their business. In 2018, Microsoft purchased GitLab, the great public repository and foundation of open-source development, for more than $7 billion in 2018, further highlighting the maturity of open-source development as a sound business strategy. In 2019, IBM followed with the purchase of Red Hat, an open-source soft-ware company, for $34 billion.

Open-source code repositories are a form of a knowledge utility necessary to support the internet. But many other types of knowledge utilities are available and equally important. These shared repositories are used because the process needed to develop and support modern systems requires very complex relationships, transactions, and modes of interoperability across communities. Open-source collaboration can do these things. In contrast, proprietary systems struggle with increased complexity and often become prohibitively expensive to maintain and enforce regardless of the size of the entity attempting to control the system. The appeal of open-source collaboration is obvious when we consider the analogy of cars and roads: Why would anyone who builds new cars also want to engineer, procure, and build new road systems to accommodate their vehicles? Instead, wouldn't it better if we shared the specifications and maps for the roads we need and spent our collective energy and creativity on building better ways to use those road, bridges, and intersections? Rather than being a radical idea—which it seemed to be a decade ago—open source is now the foundation of nearly all modern technical infrastructure, from servers and automotive software to phones and browsers.

It's not just about code. Open source is a culture that embraces the tension of collaboration and autonomy as an approach to problem-solving. The same cultural approach that has worked for software applies more broadly to research and development within public science communities. The Human Genome Project, the largest collaborative biological research endeavor in history, is a classic example of a knowledge utility in action. The Human Genome Project shares data using the underlying infrastructure of the internet but operates that backbone as a curated, trusted, and distinct knowledge utility. Begun in 1990s, the original objective of the Human Genome Project was an ambitious one: identify and publish all the DNA base pairs that comprise the human genome, and in so doing create a comprehensive resource for scientific researchers. Scientists suspected that mapping the genome would lead to breakthroughs in medical research and treatments for patients, as well as advances in agriculture, anthropology, and the study of evolution. The job was a highly complex one, requiring the coordination of an international consortium of government agencies, universities, labs, and researchers. The project was publicly funded, with the goal of making the data available for free over the internet. Project directors, however, soon found themselves in a race with a private company that hoped to complete the sequencing first and patent parts of the genome for proprietary use and profit. The public project won the race. The complete human genome was published in 2003 in a user-friendly format employing open-source software. Anyone can access the database. Researchers can build on each other's work, share data and conclusions, and develop new investigative tools collaboratively, all of which has resulted in unexpected discoveries across disciplines, profoundly changing the fields of biology and medicine.

The famed Large Hadron Collider (LHC) is an even larger example of an open-source public science project intent on unlocking the secrets of the universe. The LHC is the most powerful accelerator in the world. Housed in Switzerland and operated by CERN (Conseil Européen pour la Recherche Nucléaire), many consider the collider to be the largest public science project in history, and work on it was partly responsible for creating the vision for a modern internet. While at CERN in 1989, Sir Tim Berners-Lee created the foundational architecture of the World Wide Web, for which he was named one of *TIME* magazine's "100 Most Important People of the 20th Century." He now chairs the World Wide Web Consortium (W3C), the international standards forum for technical development of the web, and the Web Foundation, whose mission is to ensure that the World Wide Web serves humanity. He was also recognized with a Turing Award, which is considered the "Nobel Prize of Computing," for his work laying the foundation for open-source approaches to software development and communications.

Collaboration and data sharing via open-source software and the internet provide the foundational components of a functioning knowledge utility. The Human Genome Project and work at CERN represent early and successful demonstrations of collaborative technology that can be applied to complex systems, examples that can be applied to other projects whether they are biological, ecological, or agricultural. However, such projects might be seen as small-scale practice runs for the main event, which requires collaboration and public science on an even grander scale to address management and stewardship of the global climate. This progress can be further accelerated with the compounding effects of increases in computing power, data accessibility, communication, diversity, and openness as more of our knowledge, now sequestered, is made commonly available through knowledge utilities.

Looking Back to See Forward

The knowledge utility is not a new idea. Its roots go back to the "free thinkers" of mid-eighteenth century Europe and their efforts to create a knowledge commons that included everyone, not just the elites (although "everyone" was not very inclusive by today's standards). A key pioneer and prototype for much of our modern open-source work, and an example of a primordial knowledge utility, is the famous *Encyclopédie, ou dictionnaire raisonné des sciences, des arts et des métiers* (*Encyclopedia, or a Systematic Dictionary of the Sciences, Arts, and Crafts*). Published in France between 1751 and 1772, the aim of the *Encyclopédie*, in the words of its indefatigable editor Denis Diderot, was "to assemble all the knowledge scattered on the surface of the Earth . . . and to transmit it to the people who will come after us so that the works of centuries past is not useless to the centuries which follow." In other words, Diderot was making the case for creating a knowledge utility to grow our shared commonwealth.

The *Encyclopédie* was part of a longer human trajectory of cataloging knowledge stretching back to the Roman historian Pliny the Elder, whose *Historia naturalis (Natural History)* gathered everything known at the time about subjects connected to nature—and is our sole source of knowledge about some aspects of ancient Roman life. Comprising thirty-seven books, Pliny's sprawling work covers geology, anthropology, zoology, botany, astronomy, technology, agriculture, medicine, artistic endeavors, and more. Its vast scope, its comprehensive index, and its use of original sources—Pliny claims to have gleaned 20,000 facts from 2,000 books and 100 authors—set the model for subsequent encyclopedias, including Francis Bacon's pioneering effort in the early seventeenth century to organize all empirical inquiry into what he called "the branches of the tree of knowledge."

In 1728, journalist Ephraim Chambers gave this tradition a boost when his *Cyclopaedia* was published in England. Containing a wide selection of topics and diagrams, it was one of the first general encyclopedias to be published in the English language. Its popularity inspired a group of French publishers, led by Andre Le Breton, to band together to publish a similar work. The project failed to move forward, however, until Breton assigned two employees to take charge, the respected mathematician Jean d'Alembert and a young writer of little renown but abundant energy and intellect named Denis Diderot. Almost immediately, they set out to expand the project far beyond its original scope. In fact, Diderot had a subversive goal in mind for the project. He had been working as a translator and editor while authoring anonymous essays and books that challenged political and religious conventions, as well as social mores (one popular work was deemed pornographic by censors). Despite being thrown into jail for his provocations and free-thinking ways, Diderot emerged unbowed. He was determined to infuse the *Encyclopédie* with humanist philosophy, political thinking, and the latest advances in science and technology, including a huge amount of practical information, defying powerful religious and royal orthodoxies about how the world worked.

Diderot and d'Alembert employed three innovative strategies in the *Encyclopédie*. First, after scouring all the dictionaries and other resource material they could find, they amassed 74,000 topics—all arranged alphabetically. It was a subtly subversive decision. In a tradition-bound nation like France, where royal, aristocratic, and religious values were highly segregated from the rest of society, especially its trade and working classes, placing an article about a revered belief of Catholicism next to instructions on how to cure constipation was sure to raise eyebrows. That was the point: the *Encyclopédie* equalized every topic, no matter how sacred or mundane, which was quite controversial at the time.

Second, the editors developed a sophisticated and sly system of cross-references between topics in order to guide readers to related subject matter. Altogether, nearly a third of the articles had at least one cross-reference and some had as many as six, for a grand total of nearly 62,000. Many links were seditious and satirical. Among the most notorious were the cross-references from the article on Cannibals to the articles on Communion and Eucharist. It annoyed the authorities but delighted readers. The cross-references were subversive in another way. Most dictionaries communicated their information in a linear manner, creating singular visions of the truth. By using cross-references, the editors created a layered, dynamic view of knowledge that highlighted poorly understood or previously unconsidered relationships among disciplines, putting them into a kind of dialogue. In doing so, the editors wanted readers to challenge dogmas and received knowledge about the world—to think for themselves, in other words. As Diderot explained it, his goal was to collapse the "mud edifice" of baseless claims into "a vain heap of dust."

The third strategy that Diderot and d'Alembert employed to accomplish their goals was commissioning 150 contributors, including scientists, doctors, poets, playwrights, historians, linguists, philosophers, craftsmen, and amateur enthusiasts to write for the *Encyclopédie*. Voltaire and Jean-Jacques Rousseau contributed articles. André Michaux, King Louis XV's naturalist, wrote nearly 1,000 articles on minerals, plants, and animal life. The scholar Louis de Jaucourt wrote nearly 8,000 articles out of the total of 71,818. Pressed by circumstances, Diderot himself wrote thousands of articles for the first two volumes, ranging across topics as diverse as childbirth, mythology, gardening, architecture, geography, and literature. Many artists and engravers were commissioned to illustrate various craft and manufacturing processes as part of the editors' intention to place the skills and labor of working classes alongside the efforts of academics, lawyers, clerics, and other professionals as being equally important to civilization.

The *Encyclopédie* encompassed twenty million words spread out over seventeen volumes of text and eleven volumes of illustrations. It took twenty-five years to complete and nearly failed on numerous occasions. The king of France banned further publication twice (he was talked out of the first attempt by a mistress), and Pope Clement XIII strenuously condemned the entire project. The editors and their friends were constantly harassed by censors and critics, frequently requiring them to smuggle pages out of the country to get printed. Authorities repeatedly threatened Diderot with imprisonment and literal gates and bars to restrict access to the *Encyclopédie* because of the threat it posed to church and nation-state. He was saved by a combination of personal wiles, friends in high places, and the popularity of the volumes, which grew as each one was published. It wasn't simply Diderot's political or moral views, or the subversive text that he slipped into the *Encyclopédie*, that endangered his liberty. The act of collecting and distributing knowledge itself was seen as seditious to entrenched powers. Knowledge, once the domain of elites and other gatekeepers, could now be widely shared and studied thanks to Diderot and his compatriots.

Free-thinking and rational discourse were important, but so was the simple act of providing access to information that had previously been unavailable or deliberately hidden away.

If these elements sound like the wiki model of modern, open-source encyclopedias that we use today, that's because they are, but with a key difference: the *Encyclopédie* focused deliberately on the *utility* of the knowledge it contained. In its emphasis on documenting the trades, agriculture, and technology, the *Encyclopédie* was more equivalent to the curated video archives on how-to sites such as iFixit and Instructables and YouTube. These are DIY videos that instruct viewers in the core trades, crafts, and skills that enable a functioning economy. I find inspiration in the fact that the boundaries we feel we might be pushing now were pushed by others before us with as much or greater passion. Our predecessors performed this service at great risk to themselves in order to move forward a large-scale, collaborative vision of science, reason, justice, and humanity. Their work places any current challenges we face into a much broader context, which also makes current ambitious efforts seem far more plausible given the contemporary advantages and privileges at our disposal.

With the advent of the internet, entities like Wikipedia took on a role similar to Diderot's master project. Today, a wealth of information can be contained and distributed on an SD card the size of a fingernail. Yet, there is a central idea within the *Encyclopédie* that has yet to fully germinate in our modern context: A knowledge utility must combine science with useful action. Tools must not be relegated only to professional scientists but rather be made accessible and useful to all. To gain the full value of our utility, our knowledge must be widely shared and made useful. As we look at our current cacophony of collaborative communication tools such as Facebook, Instagram, Google, YouTube, and Twitter—which emerged as byproducts of a capitalist system rather than as a primary and intentional output of the economy—it is useful to contemplate the context within which the original "cyclopaedia" emerged from structured letter writing and how its genesis was used to create new sources of power based on information often at odds with official dogmas. As we witness challenges to truth, science, and reason being mounted in furious ways across our social media landscape today, with "fake news" or the selective editing of scientific findings in government reports, we can see echoes of the struggle tackled 300 years ago, albeit then with printing presses and flyers rather than the giant digital copy machine that is the current internet. At the root of Diderot's *Encyclopédie* is an idea that is still necessary and relevant today: germinating new ways to create revolutionary collaborative projects that redirect establishment power toward goals that change how we think.

There is a darker, though no less important lesson to learn from the *Encyclopédie*. Its emphasis on systems thinking was rooted in documentation as well as economics, with the goal of improving and restoring the health of the land. This goal came under fierce assault during the industrial revolution and the subsequent growth of colonial capitalism. Land-based economists and agrarian philosophers were overrun by the pace and power unlocked by the industrial application of reductionist science and sidelined in a world of specialization, which formed a repeating pattern of tensions between industry and agrarianism still playing out in the United States and internationally. Diderot's vision was buried by a new type of economy that was able to rapidly exploit scientific knowledge for extraction and depletion rather than regeneration. Knowledge became a commodity to be protected by patents, institutions, and the professionalization of science. Its purpose was pulled hard away from collective innovation and sharing and used instead to build walls that deliberately obscured visibility into our shared commonwealth of knowledge.

Although the enlightened work of the *Encyclopédie* was eclipsed by industrialism and corporate capitalism, it wrote an aspirational promissory note to future generations. As a precursor to modern knowledge utilities, its influence can be seen in the founding of the US Department of Agriculture in 1862 as well as the founding of the International Institute of Agriculture in 1905, both of which were charged with the goal of documenting, mapping, and sharing agricultural knowledge. It can also be seen in the founding of the Food and Agriculture Organization (FAO) of the United Nations, which was created following the end of World War II to assist newly decolonized nations with development of their agricultural land. Often led by postcolonial Indian economists and social scientists, the FAO set out hugely ambitious plans for disseminating seeds, creating participatory soil maps, and distributing calculators, pamphlets, and bibliographies as a collaborative system of information designed to fight poverty. Although the implementation of the early ideals was not fully realized during the first decades of the FAO, which was plagued by political and technical hurdles, it kept alive a vision for a global agricultural knowledge utility that could tap the power and potential of a shared commonwealth of agricultural knowledge.

That vision is now within our grasp.

CHAPTER 3

Public Science and Soil Health

Humankind is nature becoming self-conscious.
—ÉLISÉE RECLUS, *L'Homme et la Terre* (1905)

Building an agricultural commons is not enough, of course. Agricultural knowledge must then be applied to growing and improving the commonwealth of nature. The first director general of the UN Food and Agriculture Organization (FOA), John Boyd Orr, stated "the people wanted bread but were given statistics."[1] The ambition of the agency was data and knowledge transformed into action and development, but that ambition was not fully realized.

Successful examples extend across human history of people sharing agricultural knowledge to improve the commonwealth of nature in practice. For example, between 2,000 and 3,000 years ago the people of the Amazon Basin practiced a hybrid of foraging, forestry, and small-plot agriculture. Evidence of their stewardship remains in the dark, carbon-rich, multi-meter-deep soils they left behind, which are still productive thousands of years later. These *terra preta* soils, which were heavily modified by humans, feature a mix of charcoal, bone, compost, and manure that together increase biological activity and reduce nutrient leaching. These soils and their productivity persist and still dot the region to this day, marking the many generations of human stewardship and cultural practices on the land. These soils are an example of knowledge applied to action that continues to enrich the commonwealth of nature in the region over thousands of years, even as the original peoples and their cultural knowledge have been lost.

These ancient riches form our knowledge commonwealth, which is being rediscovered and folded into our shared understanding of land and its stewardship. Today, the USDA and organizations across the globe, such as International Biochar Initiative, as well as private enterprise, are collectively investing billions into understanding biochar, the process of creating new high-carbon soil amendments by combusting plants in low-oxygen environments, combining that with bioactive organic amendments, and applying the residue back to soils.

In this chapter we will test the concept of a knowledge utility across generations and will ground our knowledge, and in so doing unlock not just an understanding of the way the world works from the microbiome to the biosphere but also generate a shared history and a new story of our land that has not yet been told. To tell this new story, we have powerful tools at our disposal, from communications satellites, orbiting telescopes, and electron scanning microscopes to globally distributed and ubiquitous sensor networks, artificial intelligence that help process feedback from our shared nervous system, and the ability to look back in time geologically and culturally.

And yet, tools are not useful unless they are used. A gas chromatograph that analyzes chemical elements is not useful if it lacks electrical power or a skilled operator. So we must look at how these tools work not in isolation but as part of a complex ecosystem. To explore this ecosystem, let us find areas where the theory is being put into practice and where we can see the emergence of a new knowledge utility that democratizes access to parts of our world that have been previously invisible.

Let's begin by zooming out again to 90,000 feet where we can see the curvature of the Earth and the depth of the pale blue atmosphere over the horizon highlighted against the deep darkness of space. At this altitude, let's travel across the Earth by soaring out to the rocky Atlantic coastline on the eastern edge of the North American continent. From there we zero in on a dairy farm located on the edge of Casco Bay, just outside Freeport, Maine. This farm, located on the campus of Wolfe's Neck Center for Agriculture and the Environment, is milking sixty cows as well as managing diverse grazing, vegetables, perennials, and forestry systems that produce food and recreation for the surrounding community and more than 30,000 annual visitors. The 600 acres under management are also part of fresh- and saltwater systems within the marine boundary of active shellfisheries and aquaculture just off the coast.

The recorded human history on this stretch of land and coast stretches back to the Androscoggin and Kennebec people who settled the region thousands of years ago and continue to live in Maine. They were the first people to practice agriculture among the rich and varied soils exposed when the glaciers retreated following the end of the last Ice Age. Androscoggin in the eastern Abenaki language means "rocky flats flow," which refers to areas that were fished and supported abundant salmon runs. The early British colonies linked local commerce with regional and global trading practices focused on commodities, such as salt and hay, that were transported as far away as Manhattan Island. The coastal landscape represents an intersection of freshwater, saltwater, forestry, and agriculture with rural, urban, and global influences affecting the landscape and people. A farm for nearly sixty years, Wolfe's Neck Center has become home to an oceanfront campground, a demonstration farm, and an educational resource center for innovative practices in regenerative agriculture. It is also where I have served as research director for the last several years.

In the 1980s, Wolfe's Neck Center became one of the first farms in the country to pilot agricultural conservation easements to ensure stewardship across generations and has continued to explore innovative approaches to land management. The Center currently sells its milk to the Stonyfield corporation, which in turn uses it for organic yogurt production. Wolfe's Neck and Stonyfield have partnered to track soil health improvements on the land as part of a science-based goal of reducing carbon emissions. A quick look at the tools and knowledge that Wolfe's Neck farmers are using reveals an array of free and open-source software management tools—knowledge utilities—to track and share grazing rotations, field inputs, and soil test results. As the public has grown more interested in soil health, I have had the honor to put theory into practice at Wolfe's Neck by expanding the research program and sharing in-depth soils and environmental and soils data with the University of New Hampshire and the USDA Climate Hubs. I've also used this platform to convene a large-scale global collaborative effort for a universally accessible agricultural data commons called OpenTEAM, or Open Technology Ecosystem for Agricultural Management. (This data commons will be covered in more detail in later chapters.)

If we travel out to the field, we can find Wolfe's Neck Center's research coordinator, Leah Puro, monitoring soil health with a handheld tool, called

Quick Carbon, that analyzes soil color to estimate organic matter composition, as well as a prototype greenhouse gas analyzer. After spending time leading research efforts in Vietnam and at Stone Barns Center in the Hudson Valley of New York, she joined the Wolfe's Neck staff and began coordinating research collaborations with OpenTEAM. Leah describes her work with OpenTEAM "as a way for skilled and curious people from different disciplines, who would not otherwise work together, to collaborate and solve shared problems and answer complex agricultural questions." Puro's confidence in the field is evident—she is just as comfortable driving a skid-steer as hosting a fifty-person Zoom call with farmers, scientists, and technologists.

Puro recently began using a free and open-source soils mobile application called Land Potential Knowledge System (LandPKS), developed by the USDA Agricultural Research Service. Soils are a key element on any farm or ranch, but they are also one of the largest variables, even within a single field. LandPKS enables anyone with a phone and a spade to discover the potential of their land and monitor changes over time. Even the inexperienced can use LandPKS to do soil identification, measure land cover, monitor soil health, manage the farm, keep records, and link to other tools within the Open-TEAM community. LandPKS represents a key knowledge utility, making global knowledge of soil locally available to anyone anywhere.

Puro is in the process of setting up a system that further builds on soil knowledge that will utilize wireless infrastructure to set paddock boundaries on her phone and set temporary field boundaries for the lead cows in the herd who wear special communications collars, and also receive cow health indicators back. She is also able to access and share past field history and ongoing activities on her phone or computer with a free open-source program called farmOS, which works in conjunction with another application called PastureMap to plan and track the movements of the livestock around the farm as they graze through the paddocks. Puro and the Wolf's Neck farm team are using these tools to make decisions about which annual and perennial crops to plant together to improve soil health and communicate this data to interested parties. In so doing, they are linking to a broader community all over the world. As Leah tests soil and monitors research projects in collaboration with local food producers in Freeport, she meanwhile coordinates with farms across New England and south as far as the Chesapeake Bay by way of the Northeast Healthy Soil Network. The underlying processes she uses to describe her work are specific to the history of the place and people and also tell a story of global knowledge and local agriculture.

To illustrate this concept further, let's zoom back out to 90,000 feet and spin the Earth east across the Atlantic and down into the southern hemisphere, zooming to Africa and the lush and fertile Zomba highlands of Malawi. We move along the Shire River and the deeply incised Domasi Valley, where the Asian and African continental shelves meet, extending north through the headwaters of the Nile into the Red Sea. The plateau's geology was largely formed during the late Jurassic and Cretaceous Periods, roughly 150 to 65 million years ago, and later cultivated by the Chewa and the Lomwe Bantu people. In the nineteenth century, the Yao people migrated from the Bangladesh coast, followed by colonial agriculture and railroads that were established to support exports in the late nineteenth century. This development led to export-oriented plantations that persist to this day, alongside local agricultural production by descendants of these groups. Like the coast of Maine, the Zomba Plateau is also known for natural beauty and recreation.

Here, we meet Wezi / Ozzie' Mazungu, a technical assistant and farmer who has worked for more than fifteen years building a network of 600 farms encompassing 1,000 fields that work together to improve soil health. Her farm uses perennial legume/maize intercropping. Since there is not a lot of livestock in Malawi, she has been using legumes, such as cowpea and pigeon pea, to improve soil. She has been replacing fields of single crops, such as corn, with a diverse blend of crops, such as soybeans or peanuts, along with popular cereal crops to help increase yields, improve the health of the soil, and create a more resilient local food production system. In 2008, she met Sieg Snapp, who runs the Michigan State University's Global Change Learning Lab in Sub-Saharan Africa, and together they developed participatory research methods that enable large-scale collaboration and knowledge exchange. When Ms. Mazungu was in the field during the summer of 2019, she collected soil samples and crop records using the same set of free software tools to measure soil health—Quick Carbon, farmOS, and LandPKS—that Leah uses thousands of miles away on the Maine coast. These tools focus on two questions that keep coming up: What works and where? Not theoretically, but practically. And not what works industrially, but regeneratively. What practices build soil carbon and improve water cycles and water-holding capacity in the land? What native plants are best suited for the area? How might traditional knowledge be effectively implemented today?

These examples illustrate the potential for using local experiences to build a global exchange. Although Puro and Ms. Mazungu have never met, their efforts are linked through the contributions they have each made. By using and improving the same tools that access shared collective knowledge, they can adapt their management to local conditions, draw on the experiences of others to improve the results of their farms. And by using and contributing to tools that are part of a knowledge utility, each can share within their own communities while also accessing and contributing on a global scale.

Let us zoom back out and spin the globe past Asia, across the Pacific, past the Rocky Mountains until we are over the Great Plains of the United States, and then zoom down to a farm in Nebraska. More than sixty-six million years ago, the Great Plains were covered by a shallow inland sea. When the sea receded, it left behind thick marine deposits and a relatively flat terrain that became favorable to the evolution of grasslands. Over millennia, forests declined and grasslands became dominant, which provided an ecological niche for mammoths, saber-toothed cats, giant sloths, horses, mastodons, and the American lion. And when these animals declined, the great herds of bison took their place. The many generations of ancestors of the Omaha, Missouri, Ponca, Pawnee, Otoe, various branches of the Lakota (Sioux) peoples, and others managed the grasslands for thousands of years before European colonists arrived in the 1600s. Millions of years of geology and thousands of years of grasslands created the foundation for a region that transitioned from grasslands to industrial cultivation in the span of a mere 300 years.

We now zoom to Knuth Farms, where Kerry and Angela Knuth, with their sons Gregory and Garrison, manage 2,200 acres. As they work to improve their soil's health by minimizing tillage, managing multiple crop rotations, and introducing diversity into their cover cropping, they are also reintegrating livestock into their fields of corn, soybeans, wheat, alfalfa, oats, cereal rye grass, and sorghum-Sudan grass. In their hometown of Mead, in Saunders County, much has changed in the past 200 years, including the condition of the land the Knuths inherited from the previous generation in the form of depleted soils, high rates of erosion, herbicide-resistant weeds, disease, drought, and flooding, all of which presented great challenges. Moreover, farmers in their area have been battling rising costs of production, amplified by the effects of extreme weather events. Recognizing the potential for regenerative agriculture to reduce their dependence on synthetic inputs and to lower equipment costs, Knuth Farms has made a significant change in their practices, adapting the tools they need to meet the challenge.

The Knuths are transitioning to an agriculture based on improving soil and biological function. However, they are no strangers to technology. In their quest to regenerate soil and restore biodiversity, they use satellite imagery, tractor guidance systems, on-farm communications tools that link grain trucks with combines, and complex software packages typical of precision commodity agriculture that are now being connected through free and open-source tools. This open-source software bridges conventional AgTech with new tools to support collecting and sharing soils data, crop planning, and feedback systems through the same survey tools that were field tested in very different conditions in Malawi and in Maine. Each grows our commonwealth of agricultural knowledge. With Angela Knuth's comprehensive tracking of verifiable, historical soil management and cropping practices for at least three years prior, she was able to engage with newly formed private carbon markets and get paid to restore soil health and provide environmental services.

We can see with Angela Knuth's work that while her farm's production scale, local markets, and motivations are vastly different from Leah Puro's or Ms. Mazungu's, the underlying questions the Knuths have are similar, and the Knuths utilize the same tools to make decisions about what to plant where and when based on soils, climate, markets, and available equipment. As we look at the digital tools the Knuths are using, we can see knowledge utilities that are a direct descendant of the clay tablets of Mesopotamia of 1800 BCE, the *Encyclopédie* in the 1700s, the land grant colleges of the 1800s, and the FAO archives and pamphlets of the twentieth century. These new knowledge utilities of the Knuths not only share some of the same computer code used by Puro and Ms. Mazungu but also the same satellite systems, the same way of describing soil health, and the same way of tracking how water moves through the soil. Puro and the Knuths share the same global pool of nitrogen, carbon, and water vapor that their plants and animals capture from the atmosphere, yet each observes a unique expression of that biology. Each is a land steward within their domain and each is creating new nutrient and water cycles that extend beyond their field edges into surrounding forests, rivers, and shorelines.

Let us now zoom out to look at another production system, this time in the Southwest, not too far from Nebraska, although it could just as well be in Patagonia or the Australian Outback. Let us travel to where the mountains meet the plains in northeastern New Mexico. It's grass country, visited over the centuries by herds of migratory bison and managed by bands of nomadic Ute, Comanche, and Apache peoples for thousands of years. European immigrants to North America began arriving in the 1700s, in addition to Hispanic homesteaders from Mexico and from communities along the Rio Grande Valley to the west. By the 1840s, trade on the Santa Fe Trail was active. A pile of discarded frontier wagons gave a nearby town its name—Wagon Mound. The entire area is known for its long vistas, deep canyons, and abundance of cool- and warm-season grasses that were later grazed by the colonizers' cattle operations.

The 13,000 acres now known as Sol Ranch is a cow-calf operation run by Emily Cornell, a young rancher who is in the process of taking over the ranch from her parents. This family markets their grass-finished beef through locally owned farm stands as well as directly to customers. Cornell's parents employed holistic grazing principles on the ranch and strove to continuously improve the health of the soil and the land, fostering biologically diverse ecosystems, and Emily is building on this approach. A theme in this landscape, and one overarching concern for Cornell, is water. Droughts have become more common in the area, as have destructive floods, both weather extremes exacerbated by climate change. Managing water, keeping it on the land, and getting it to productively infiltrate into the soil—all while dealing with periodic floods and threats of erosion—has become a big part of modern ranch work.

Cornell left the ranch for a while when she was younger and returned with the skills of a researcher and range technician, including a familiarity with technology and the ability to set up data loggers. She worked with her father to gain a deeper understanding of holistic management in order to create a five-year framework for managing the ranch according to testable goals and scientific monitoring of land conditions. The job is daunting with so many choices to be made, each one dependent on ecological variables that can rise or fall dramatically each year. She quickly found that interpreting the large amounts of information and translating it into decisions and actions required a public science team effort.

Quivira Coalition, a nonprofit organization based in Santa Fe, met this need with their newly formed Carbon Ranch Initiative program, which helps land stewards improve soil health through a knowledge sharing and support network. The Quivira Coalition is also a member of OpenTEAM. New Mexico State University has also stepped in to help map soils and vegetation across the ranch, here again using LandPKS to access a global library of networked knowledge utilities that combine hardware and software to document observable local changes over time. But LandPKS is not the only link to global knowledge utilities. On Sol Ranch, Cornell tracks management decisions as well as weather events and their effects on the physical conditions of the ranch. For her cattle management, she is using a software program adapted from Australia, similar to farmOS and PastureMap, to track livestock management, including paddock size, grazing periods, stocking rates, forage conditions, available nutrition, animal health, reproduction rates, and economic projections.

Cornell's interest in improving resilience in her operation led her to use her research background to reach out and collaborate with others. She established experimental plots on the ranch where grasses are protected from grazing pressure, and collected data on how to manage water on the ranch more efficiently by mapping the invisible rivers through hydrographs of underground water flow and storage. These extrasensory tools form the foundation for better planning as the well-water recharge process becomes visible to her and to others. It's a critical issue for land stewards anywhere in dry country, and one that will be ever more important with the increasingly extreme weather already being experienced in a rapidly changing climate.

Cornell sees Sol Ranch as both a livestock operation and a lifelong research project. The key is making the invisible water and carbon cycles visible, which is now possible due to the sophisticated, affordable hardware, software, and large-scale collaboration. "It's hard to do alone," she told me. "Collaboration is absolutely required, whether it's a nonprofit, a scientist, or your neighbor. To tackle all the complexity today involving animals, land, or data and to move the big issues forward, you need help."

As we zoom back out we can again see the context of millions of other land stewards that now have access to and can contribute to the same knowledge base that these four land stewards do.

Throwing Off Blinders

Our journey toward a shared knowledge commons flows well beyond the story of land and land stewards. The process that makes the invisible visible is also a process of throwing off our social and cultural blinders so we might directly see the hopeful stories of the land that supports us, and also the not-so-beautiful stories that continue to surround us. These stories must also be part of the solution. While scientific knowledge was beginning to illuminate what was historically invisible, other industrial and powerful forces were and still are working in reverse.

Much of the industrial use of technology—as well as complex organizational and legal structures, academic admissions, paywalls, zoning and building codes, and other forms of barriers—are designed to limit our field of vision and hide the pain and suffering of nature, people, and communities involved in the creation of industrial goods. Many of these "innovations" were made possible by concealing moral quandaries that exist partly because those quandaries can be put out of mind by being out of sight. This has been especially true across the history of agriculture. This is not news. The legacy of environmental degradation and "invisible" labor—slavery, forced labor, child labor, and undocumented and/or low-wage workers—in the production of food and fiber is undeniable not only for the history of the United States but throughout recorded history: from feudal China, the Roman Empire, and ancient Egypt to more recent examples of colonial industrial export commodity production. It is not the only human story, but it is a consistent one when agriculture is combined with scarcity. Classic books about exploitation of land and people all seem to have a similar theme in bringing to light what was intentionally being made invisible, such as Upton Sinclair's *The Jungle* (meatpacking industry); John Steinbeck's *The Grapes of Wrath* (farm labor); Rachel Carson's *Silent Spring* (chemical pesticides); David Montgomery's *Dirt: The Erosion of Civilizations* (soil degradation); Leah Penniman's *Farming While Black* (racial injustice in agriculture); and a long stream of documentaries about sweat shops, illegal mines, forced-labor electronics assembly, and more. As we make our land and people more visible, such as the four farms we just explored in this chapter, we also elevate individuals and groups and make it harder for aspects of our economy to be out of sight or out of mind.

When I worked in Hong Kong in 1999, I toured a leather plant in southern China run by a Brazilian company. In the middle of a sprawling complex of drab concrete surrounded by walls and barbed wire sat pale-blue cinderblock bunkers for workers. Also within this complex was a different kind of structure. This one had bright white, high stucco walls, topped with terracotta roof tiles. The ornate entrance had the feel of a country club gate, intended to be welcoming to some and exclusionary to others. Once inside, the stark gray concrete of that industrial world outside disappeared and melted into an expansive bright green lawn and a pastoral picnic party scene with kids playing soccer and laughing and adults gathering around the assado set up to slow roast meat in the traditional Brazilian fashion. This aspect of the tour was presented as a point of pride, in that it demonstrated the company could provide a sense of home for the Brazilian employees and their families in the midst of an industrial city in southern China. The bleakness of the prison-like settings of these two walled communities, one within the other, stuck with me as an example of how our technology is frequently used to deliberately hide things it (or we) do not want to see. It enables us to agree to things that we might never agree to if we could see them in our own community. In fact, one of the core innovations of colonialism seems to be social technology (such as bureaucratic specialization, chains of command, and professional credential–based access to knowledge) that has narrowed our fields of vision, making critical parts of our world invisible. The tools we create are an embodiment of our values regardless of any positive and negative environmental side effects. As Langdon Winner said, tools also "can embody specific forms of power and authority."[2]

Colonialism, followed by modern industrialism, wasn't just about efficiency in extraction but also about using these same technologies to limit the available field of view as a form of control. The advancement of knowledge through industrialization erred toward restricting access through professionalization, legal protections, and the specialization of science, more than through the democratization and distribution of knowledge. The public discussion of a "digital divide," particularly for rural communities, has largely been framed as a political issue of access to "rural broadband." However, the larger issue is access to critical infrastructure as well as the opportunities associated with access to knowledge-based economies and markets. At issue are the available communications tools and utilities that can limit available perspectives on the world whether we realize it or not.

In 1971, the media theorist Marshall McLuhan speculated that "every home is going to have a computer and you will have access to all the films, and all the entertainment and all the information that ever was." Pete Townshend, cofounder of The Who and an electronic music pioneer, took inspiration from this idea for his Lifehouse project, an unfinished rock opera that imagined a dystopian world in which everyone thought they had access to all available knowledge because of the perceived richness and sheer number of choices, but in reality the knowledge fed to everyone on the "grid" was carefully curated and incomplete. The central control of an incomplete knowledge utility is indeed something to be vigilant against.

The current struggle with inequity as it pertains to remote learning echoes historical concerns about the distribution of printing presses, the direction of roads and rail lines across a landscape, the rate of rural electrification, the placement of radio, TV, and cellular towers—and now satellite, data towers, and fiber optic routes. Even as the internet was proposed as a democratizing commons, it became dominated by centralized power structures and corporations, which echoes enclosure movements of the past.

Technology embodies the values and worldview of those creating it. A carriage horse's bridle is a form of technology which uses blinders to enable greater efficiency; blocking out the distraction of the rest of the world enables the horse to concentrate on its task, to stay focused solely on the road ahead. However, the life of a carriage horse is certainly not one to aspire to. This was very much Diderot's aim as he produced the *Encyclopédie*—to throw off the dominant society's blinders. Inspired by the representative governance of Native American peoples encountered in Western colonies, the coffee shops and salons in prerevolutionary France fostered a diversity of thought and natural philosophy that challenged the assumptions of the power of church and hereditary monarchy. North American colonizers encountered systems of governance that were not based on rigid castes. These more democratic social structures were far from the reach of the power of a crown. The mere existence of an alternative to the church and hereditary monarchy was a direct challenge to power and central authority and was a key pattern that lead toward what would also enable the unblinkering of modern public science. From this longer view we can see our current context as midstream in an exciting and dynamic story that is still unfolding new knowledge architectures and utilities that build on the unfinished work of our predecessors.

I see this vision made manifest when I join an OpenTEAM call nearly every day. These gatherings can have participants from five time zones and include academics, software developers, and representatives from food companies and farmer networks from around the world. All are coming together to increase our shared commonwealth of knowledge and improve our shared commonwealth of nature with our stewardship. When I pour my own cup of coffee in the morning before heading out to check water for the grazing sheep, I like to think of folks also rising to care for the land in Nebraska, Malawi, New Mexico, and beyond following a similar pattern. I like to think of the origin of all the others on the land before me who gathered over coffee to think and converse as they contemplated alternatives to the status quo. If we look at the longer trajectory of logic and reason, we can see a universal human process of asking questions and seeking answers over all manner of shared brewed beverages.

CHAPTER 4

Farm Hacks and Open-Source Observatories

We become what we behold. We shape our tools and then our tools shape us.
—MARSHALL MCLUHAN & JOHN CULKIN

In 2011, as I watched a purple weather balloon rise from a farm field in New Hampshire, I gained a new understanding of the concept of a knowledge utility. With that balloon, a new kind of observatory was available. I watched it with a volunteer group from Public Lab who had driven up for the day from Cambridge, Massachusetts. Public Lab is a citizen science organization founded in the wake of the 2010 Gulf of Mexico oil spill to help document and map the effects of the spill. I got to know this skilled amateur science group through my connections at MIT. The Public Lab community expanded on the knowledge utility idea by providing and documenting low-cost, DIY scientific tools to help understand the world. This work would subsequently inform the evolution of Farm Hack (covered later in this chapter). Both communities helped illustrate to me the power of a shared agricultural knowledge utility and prompted me to think about the importance of observatories in our lives. An observatory is a place or institution designed and equipped for making observations of natural phenomena, like a structure overlooking an extensive view. An observatory represents a particular vision for public science that is distinct from the vision of white-coated scientists in labs. It represents shared exploration as much as individual experimentation.

In 2011, I was conducting no-till cover crop trials in a field outside the University of New Hampshire as part of my graduate work. I wanted to digitally photograph the field to capture the dramatic effects of biochar and wood ash amendments on legume plant growth in a way that would communicate what I was seeing with more nuance than charts and tables could provide. I thought an overhead digital image would do the trick to show the dramatic difference in plant growth and the density of green in the various plots. At the time, however, even a friend's radio-controlled helicopter didn't have the capacity to carry a digital camera of sufficient quality (this was before the proliferation of commercial multirotor drones or even GoPro cameras).

As the purple balloon rose, I saw the potential for a new system that could leap past proprietary industrial agricultural technology. The balloon we sent up carried two thirty-dollar digital, point-and-shoot Canon cameras, purchased from eBay, and modified with hacked firmware that, through a simple script loaded from the SD card, enabled the cameras to take images at predetermined intervals to create a digital map of the field. In addition, the Public Lab community was working with the same software used to create satellite image maps in near-infrared bands to calculate photosynthetic performance related to Normalized Differential Vegetation Index (NDVI). The images would be combined using free and open-source software using a technique called structure from motion, which takes hundreds of still images of an object and stitches them together to create an accurate 3D model of the landscape. The result would be a high-resolution, three-dimensional map of the plant growth in the plots in the field with enough resolution to zoom down to the level of individual plant leaves. Watching the balloon, I realized that every citizen could now potentially access free and open-source tools to see their land in high resolution in a way that had previously been the exclusive domain of government agencies or corporations that could afford satellites, aircraft, and expensive proprietary technology.

The suspended camera, stabilized by a harness of kite string and 3D-printed plastic parts, symbolized a new kind of cross-pollination among researchers, with implications for social systems based on a new kind of science—an exciting public science that stirred echoes of the preindustrial, amateur-led advancement of shared knowledge. The low cost of the cameras and balloon, combined with the open-source software, meant a whole new level of participation and precision was now possible, creating not only the opportunity for a shared understanding of our environment but also a society that could democratize technological development rather than continuing to be controlled by it. What happened in the field that day was not the product of a multinational research lab, a corporation, or a venture fund–backed start-up, but the culmination of effort led by a citizen group, collaborating through online forums, to create powerful, low-cost tools that could generate *lots* of data. As with the *Encyclopédie*, the fire of ideas can spread regardless of the intent of the kindling (in this case, the underlying complementary metal-oxide semiconductor digital image chips and associated integrated circuits which originally came out of NASA). The data-collection capability for nearly everyone on Earth has exploded since 2011 when I watched that purple balloon rise into the air.

A thirty dollar digital camera running open-source Canon-hack development kit (CHDK) software is suspended from a balloon in preparation for the first flight to capture high resolution aerial imagery of farm fields at a collaborative event between Public Lab and Farm Hack communities.

Not only do these developments present important opportunities to farmers, ranchers, and everyone else connected with agriculture, they are also significant for the opposition they represent to the domination and control exercised by those who have had a monopoly on the industrial consolidation of information. Using the same digital technology to connect individuals and organizations with each other via peer-to-peer networks decentralizes power while increasing adaptive capacity and diversity. Instead of creating tools that separate us from the natural world, these tools act like new human senses in that they provide universal access to knowledge and connectivity to the world we live in and must manage together.

This access has never felt more possible or more necessary. It is our burden, as well as our privilege, to be able to contemplate the potential for both the destruction *and* the regeneration of the planetwide systems that sustain us. We know now in great detail the effects and consequences of the degeneration of natural systems, but we also can begin to use the tools in our hands to create abundance and restore health. Every day we see the risks of failing to harness this vast power we wield with a unified purpose. If we do not choose to "take the tiller in hand," there is abundant evidence that these same powers will be used to hide truth and sow discord rather than build common trust.

———

At the end of the balloon trial in that field, I had a 32 GB SD card full of images of cover crop characteristics. Each pixel of color, each bloom or leaf, was a tile, in a sense, of a larger mosaic that represented that farm field. In art, a mosaic is a larger work composed of small, individual pieces that form a cohesive whole when observed from a distance. The ancient Romans were crazy about mosaics made from tiny colored ceramic tiles, and with great craftsmanship they decorated public and private spaces with intricate storytelling mosaics, which are still being uncovered today. In our case, the bigger pattern and story we are uncovering is about nature and agriculture. After the balloon trial, I had within my grasp the ability to elevate my perspective and look down from the third-person view at the whole mosaic of the research field—from the patterns of the plots together, down to the deep shades of greens, the black spots of wood ash, and the rich purple of the hairy vetch flowers. That same simple balloon and apparatus, filled with helium from a local grocery store, could image an individual leaf pattern at fifty feet above the field, or if released from its tether, rise tens of thousands of feet and place the field in a regional context from the edge of the atmosphere.

As the result of the vast shared work of others that I could now access, I had the ability to see both the bigger patterns of this field as well as zoom in on the details—and then share both these big and small perspectives with others. The details of the leaves, the colors of the flowers, and the volume of the biomass, among many other attributes of the living mosaic patterns were all embedded in those pixels now stored on an SD card. It occurred to me that the images were just one piece of the data needed to create a human-readable code to describe the much more complex code being expressed in the field. Each seed and fungal spore was a different kind of ancient source code that we were attempting to understand through proxy. I could start to see my work of sharing code on little silicon SD cards as similar to the way seeds exchange complex information. From this experience I began to see patterns of a greater commonwealth of knowledge in the Narragansett flint corn seed we save each year and grow, and which was developed through generations of seed saving by people I will never know. It was through this experience that I began to see that the tools we are using could be harnessed to create a digital analog to the seeds we share so easily, seeds that also encapsulate the related knowledge in genetic biodiversity that answer the simple question: "What works where?"

Farm Hack: Open-Source Agriculture

The early gatherings of Farm Hack, convened through the MIT D-Lab (the D is for Development) in Cambridge, Massachusetts, sprouted from the simple observation that managing our environment is a public science and a shared human endeavor that requires large-scale collaboration. My connection with Farm Hack started with conversations with Severine Fleming and with the newly formed National Young Farmers Coalition (NYFC), a nonprofit organization founded in 2010 in New York. Inspired by the lack of tools to serve the needs of an emerging young farmer movement, we wanted to manifest and test what an "Open-Source Community for a Resilient Agriculture" might look like. In 2011, the idea evolved from initial design sessions at MIT that brought together farmers, software engineers, roboticists, designers, artists, and food-system leaders into an online platform with the goal of documenting, sharing, improving, and fabricating farm tools. Little did we know at the time, the deep history of agrarian development theory that we were tapping into had been artfully articulated more than sixty years earlier by renowned Indian economists, including Samar Ranjan Sen (who likely also inspired E. F. Schumacher's *Small Is Beautiful*).

The contemporary update of these concepts was sparked by the successful growth of other open-source software communities such as Drupal, which were beginning to dominate certain types of web publishing. The community then began building on the capabilities of these existing open platforms to generate cross-referenced wiki-type articles and "how-to" information involving designs from around the world. The first set of tools to be documented represented the interests and needs of the users and included a wide range of small-scale manufacturing tools, small axial-flow combine harvesters from India and China, newly designed pedal-powered root washers, manufacturing blueprints for small-scale grain-hulling machines, greenhouse automation systems, soil-carbon measurement devices, and low-cost balloon-mounted camera setups.

When I first connected with Farm Hack, I had just been introduced to open-source hardware through the online biodiesel community and was partway through documenting a self-contained mobile biodiesel processor that I was fabricating. I was actively thinking through how I might best document and share the results and collaborate with a wider community. The biodiesel processor became the first tool I documented, and this project launched me into a new line of thinking about sharing and abundance, which also linked me to global postcolonial development efforts that aimed to create agrarian prosperity without dependency or debt. These ideas harkened back to my childhood bookshelves and dinner table conversations in New England and were now manifesting themselves in new and exciting ways. Farm Hack emerged as a central driving part of my identity from that point forward.

The word *hack* comes from the tech communities at the time and predates its current, more ubiquitous use. We employed it as a kind of subversive repurposing, with the goal of taking control of one's destiny. A hack is an unusual assembly of available components. A component or module is a discrete portion of a tool that has a particular function. Components can be assembled in different combinations to create hacks or tools. The larger the library of components, the more "genetic" diversity to choose from when creating new tools or hacks to address a particular challenge. A hack is an individual effort and creates an isolated workable solution. It is the basis for empowerment and innovation using global knowledge and local production. A tool is any workable hack that has been tested and replicated over time and by other parties. A tool can be a physical object, method, or framework that can be documented (software, for example). And a hackathon is a sustained effort to produce them.

Soon after its founding, the Farm Hack community grew rapidly and began generating a flurry of creative solutions through dozens of hackathons hosted in farms, grange halls, and makerspaces across the United States. These efforts built on the energy and passion of young farmers connected informally through groups like the Greenhorns and the National Young Farmers Coalition, as well as through open-source technology networks sparked by the initial MIT event. Even from the first event, there were connections to other related international projects such as the French effort called L'Atelier Paysan, which tied back to FAO agrarian development theory and focused on knowledge and small-scale technology as a path to independence for the developing world. Now these combined efforts de-eloping local technology and manufacturing capacity were being applied to foster local food economies in the developed world, too.

The first participants of Farm Hack events were not only farmers but people with common interests who had started to make the connection that improving agriculture is a shared human endeavor. The participants continued to find common ground between engineers, roboticists, designers, architects, fabricators, tinkerers, programmers, and hackers, as well as everyone participating in the wider food system. The work of Farm Hack was based on the idea that agricultural interests are a collective effort; that if we could provide a place to organize and exchange, then the ideas, inspiration, and reciprocity would flow from there. From the beginning, it was never a requirement for participants to have a farm or specialized skills to join the forum or attend an event. Farm Hack is a participatory and cumulative project that is only as strong as the exchange of ideas that flow from interactions, online or offline. And strong it is because, as we will see, Farm Hack formed the foundation of later, larger-scale efforts such as OpenTEAM.

Nine principles emerged from the early Farm Hack events that have continued to be applied to agricultural questions and design principles:

BIOLOGY BEFORE STEEL AND DIESEL: Approach problems by using biological
systems that improve soil health—such as practices like cover cropping,
mulching, crop, or animal rotations. Look to nature and tradition.

HOLISTIC APPROACH: Ask, does this tool make me enjoy working with it as
much as getting the job done faster?

UNIVERSALITY: Whenever possible, use standardized components, measurements, and systems to allow easier replication and alteration.

TRANSPARENCY: Functional components are clearly laid out and their purpose is clear.

MODULARITY: Functions can be removed and replaced without reengi-
neering the entire tool. Tool function can be changed by adding or
subtracting parts.

ADAPTABILITY: Tools can be used for many functions and can be changed
for new functions easily.

DESIGN FOR DISASSEMBLY: If welded, make the joint easy to access. Don't
hide bolts, bearings, or belts. Design belt and chain tensioners to have
enough play to enable easy removal.

REPLICABILITY: Is it well documented? Can it be manufactured locally?
Could this part be re-created in a farm shop in a small town? Does
it use common-dimension materials (for example, design for welded
and machined parts rather than castings)? Does it use off-the-shelf or
commonly available components or components that are or can be
repurposed? Can a more easily sourced part do the job as well?

AFFORDABILITY: Is this design more affordable to build than a conventional/
proprietary alternative, while still being durable and high quality?

These rules of thumb emerged at a similar time as the FAIR data principles
(Findable, Accessible, Interoperable, Reusable) promoted by organizations
such as GODAN (Global Open Data for Agriculture and Nutrition). FAIR
principles encourage institutions to serve and share knowledge rather
than to follow the usual route of locking knowledge up in obscure journals or inaccessible scientific articles. FAIR principles have subsequently
been adapted as the CARE Principles for Indigenous Data Governance

(Collective Benefit, Authority to Control, Responsibility, Ethics) were drafted in response to the UN Declaration on the Rights of Indigenous Peoples (UNDRIP), which reaffirms Indigenous rights to self-governance and authority to control their cultural heritage, and which is embedded in their languages, knowledge, practices, technologies, natural resources, and territories. FAIR and CARE principles taken together embody the tension between collaboration and sovereignty, and contradictions of creating greater mutually dependent independence.

By posting a design on Farm Hack, users are able to gain the experience of committing and contributing to open-source knowledge licensing. This means that a user's tool pages are accessible and editable by any other member of the community and can be freely built upon and shared by default using the Creative Commons license. This open approach is a direct contrast with the way the vast majority of commercial agricultural tools were being built at the time in 2011. Most often companies invest money in research and development and license their designs in a way that does not allow others to replicate it, or even know how it is made. In contrast, Farm Hack was an early open-source community that tested what a process might look like where everyone benefits from freely sharing knowledge.

From the beginning, Farm Hack operated as a community, a civically informed and independent citizenry working together toward the common good. What could generate more independence than the ability to understand and improve our agroecosystems in order to feed and fuel ourselves without reliance on proprietary interests? Farm Hack, at its core, was based on an idea about how communities can translate and share questions and knowledge through both conversations online and in person. It was a test of the core belief that we all become better farmers and better citizens when we work together. The politics of the tools created by Farm Hack focus on parts of the technical ecosystem that can be distributed, created, and managed locally rather than on the things needed to be purchased to achieve more independence.

In organizing the first Farm Hack events, it became clear that the collaborative, rather than the competitive, approach is not a battle that can be won but an ongoing endeavor of continual improvement. It is a process that is accelerated through generations of accumulated knowledge and continual improvement, first in our understanding of the environment and then in our ability to produce regenerative outcomes. A key from the beginning of Farm Hack's exploration was the idea that community was created both online and in person, and that to scale community we must each be able to extend and share conversations beyond the local coffee shop.

Farm Hack aimed to test the idea that to nurture the development, documentation, and manufacture of farm tools for resilient agriculture it would be important to also build a community of collaboration with like-minded organizations. Farm Hack struck the spark for a self-governing community that builds its own capacity and content, rather than following a traditional cycle of raising money to fund top-down knowledge generation and guide-book writing. As exciting as it was for me to have found the community I had been looking for, it was also clear that we were still in a nascent stage of development in our scope and complexity. In other words, Farm Hack was a prototype for a particular kind of knowledge utility. It illustrated the belief that greater knowledge sharing will lead to better tools, skills, and systems to build successful, resilient farms. It was a reaffirmation of the agrarian idea that open-sourcing seeds, breeds, and technology is the fastest way to accelerate innovation and adaptation as well as to ensure an equitable, diverse agricultural landscape.

Farm Hack illustrated that people can be motivated and inspired by documenting, sharing, and improving farm tools. It was clear that there was power in the idea that together we can improve the productivity and viability of sustainable farming and local manufacturing, especially when we share the goals of healthy land, abundant food, successful farm businesses, and invigorated local economies.

The Tools We Use

My experience with Farm Hack and Public Lab led to new revelations about how tools can move our ideas from theory to practice. If the tools are in our hands and used with intention, we can shape a future we want; if they are neither, the future will be shaped for us by someone else. When we think of tools, the first objects that come to mind might be familiar ones that we have created to manipulate the environment in which we live—such as a hammer, screwdriver, axe, plow—or tools that make other tools, like a milling machine. Tools are not simply objects, however. There is skill in making and wielding them, learning from them, and communicating our knowledge and experience to others through them. Embedded in each and every tool we use—from smartphones, cover crops, and encyclopedias to Zoom calls, search bars, waffle irons, and coffee makers—is the reflection of a shared understanding of the environment and the context in which they are created.

The hammer is the oldest tool in human use, dating back 30,000 years or more when stone was first bound to wood handles with animal sinew. The metal hammerhead dates to the Iron Age (and is remarkably similar to a modern design) and was well employed as civilization expanded. The Romans invented the claw hammer as a way to pull out iron nails used in the construction of thousands of buildings across the empire. Through the centuries, variations of the hammer have been developed for multiple purposes, including shoemaking, framing a house, constructing masonry, creating a slate roof, demolition, or as a crude but effective weapon. Axes, adzes, mattocks, and chisels are variations on the same tool idea defined thousands of years ago. Context is important, too. To a skilled forester, a wedge has a different meaning than to a book binder.

Beyond their physical nature, tools are also a reflection of our hopes and fears. Some people may carry binoculars or a camera with a telephoto lens into a park in hopeful anticipation of seeing a rare bird, while others may fearfully carry a gun into the same park because of dangers either imagined or real. It's the same with the tools we create to observe, analyze, and communicate knowledge. They shape us and the world we wish to regenerate. By creating them, we commit ourselves to the consequences of using them. The tools we use can build trust and empathy, or they can make us more fearful. As tools become ever more powerful through automation and networking, and less under our governance as a result, the stakes become higher. Marie Curie's statement becomes ever more salient: "Now is a time to understand more, so that we may fear less."

Observational tools extend our own senses in time and space (cameras, flashlights, magnifying glasses and telescopes, borescopes, game cameras, GPS trackers, and so forth), and become part of a much more complex dynamic than simply documenting the shape and metallurgy of a plow or the curve of a laminate bow. Observational tools, rare and rudimentary for most of history, have become ubiquitous in recent years, and with them come worries about abuse. The same camera that can be used for time-lapse imagery of a garden can be used for home surveillance. What makes a tool work is the knowledge and context that accompanies the tool—its creation, distribution, and ultimate use. What makes the matter even more interesting is that tools are not limited to physical objects. For example the knowledge of a particular radish's ability to break up compacted soil and take up surplus nitrogen can substitute for a tractor, plow, and diesel to do a similar job. And just as a tractor can be used to build or degrade soil, we can also ask in what way a radish degrades or improves the soil. We must address an essential question: How do we share our understanding of a tool and the skills necessary to best use it?

Key to the use of tools is knowledge utilities providing a two-way, shared knowledge exchange so people have the means to increase our commonwealth of knowledge and apply that knowledge to increase the commonwealth of nature. In my view, rapidly increasing the commonwealth is critical to both healing the planet and growing a regenerative economy. Just like power utility infrastructure such as transmission lines, substations, and transformers are necessary to move electricity, a similar scale of hardware and software utilities are required to support a global knowledge commons. It is confidence in this infrastructure that enables your family to plug a 110V charger into the wall and charge a 5V phone from a 460V three-phase powerline outside the house. It is indeed important to have confidence that power in the outlet is 110V rather than 460V. Similarly, we need knowledge utilities to help filter and translate relevant knowledge to a usable level as it flows to us. So what is the right combination of action, observation, and communication tools we can use that both provides trust while also unlocking innovation? The process of adaptive management requires the skilled use of all four types of tools. In chapter 7 we will look at the hardware and software infrastructure that is being built to enable just this sort of carbon–silicon nervous system.

Choosing Our Tools

Kitchens are filled with technical knowledge and highly specialized tools. They are places full of tools embedded with special context and knowledge, where people come together for the preparation of, and the sharing of stories around, food. A professional chef and an amateur have access to the same measuring cups, cutters, rollers, mashers, knives, pots, and pans, but they create entirely different meals even if they use similar ingredients. Chefs may rely on their culinary training and experience to create meals, while novice cooks may have to rely on recipes for success. The recipe, like software code, is a tool for reproducing results. Good recipes help answer the questions "What works well where?" and "What has worked well over time?" based on a set of variables—such as available ingredients, temperature, and time—which drive the biological and chemical processes of cooking. A recipe is a special kind of communication tool that becomes culturally powerful as it is shared. A recipe's effectiveness is in giving one meal maker the ability to reproduce the results of another while also enabling innovation and adaptation to local conditions. Take, for example, a recipe for making a loaf of bread. Whether it's printed on a card or in a book or blogged or produced as a video, each recipe expresses the process for making bread utilizing the same basic ingredients of flour, yeast, salt, and water, and yet the results might vary widely because it is in the details of the process that the loaf is created.

Similarly, recipe books represent another form of a knowledge utility that can also explicitly be coded into the devices we use every day. There is an unlimited number of ways to make a cup of coffee, and yet a coffee maker as a tool can also guide specific aspects of the recipe, embedding another's knowledge into the process. There is scientific and mechanical knowledge embedded in the machine that goes well beyond temperature and flow rate of the water through the grounds—the tempered glass, the heating element, any molded plastic elements, the proper amount of stainless steel, the timer, the temperature sensor, and all the microcircuitry on printed boards. The grinder used to prepare the beans is the product of a long history of experimentation to get the burr and clearance for grinding just right. It's the same with every object in your kitchen. The embedded knowledge runs deep. This concept is crucial to understand the important roles that knowledge utilities, open-source technology, and regenerative agriculture have in managing the complex systems that are needed to tackle climate change and other challenges involved in creating the Great Regeneration. The embedded knowledge in objects represent a process of coding our individual knowledge and localized recipes into the global commonwealth of knowledge. When

A Select History of Agricultural Technology

20,000+ YEARS BEFORE PRESENT (BP)	Hunting and fishing tools
12,000+ BP	Smoking method of meat and seafood preservation
11,000+ BP	Plant and animal domestication
9,000 BP	Commodity currency and tokens of debt
2,450 BP	*Terra preta* soils
2,133 BP	Lex Agraria, the Roman law proposing the redistribution of land to farmers
12TH CENTURY	Oral constitution of the Iroquois Confederacy "great law of peace"
14TH CENTURY	Dry-salting method of long-term meat preservation
1608	Refracting telescope
1701	Seed drill
1786	Threshing machine
1787	Power loom
1794	Cotton gin
1831	Commercially successful reaper
1837	Commercially successful cast-steel plow
1853	Rotary printing press
1856	British steam traction engine
1858	Aerial imagery (from a balloon)
1859	Modern oil well
1862	Abraham Lincoln's independent Department of Agriculture
1867	National Grange of the Patrons of Husbandry, a national agricultural advocacy group
1870	Stationary baler or hay press
1877	Grain warehouses a "private utility in the public interest"

1878 Successful oil tanker, the *Zoroaster*

1879 Milking machine

1881 Compression refrigeration unit for
meat shipment

1887 Hatch Act for the federal funding of agricultural
experiment stations

1888 Aerial imagery over 1,000 feet of altitude (kite)

1902 Farmers Educational and Cooperative
Union of America

1906 International Institute of Agriculture (IIA)

1907 Industrial thermoplastics (Bakelite)

1907 Commercial arc furnace

1908 Diesel patent's expirations

1911 Self-propelled combine harvester

1913 Industrial-scale Haber-Bosch process in BASF's
Oppau plant in Germany

1914 The Smith-Lever Act funding of
cooperative extension

1919 Modern techniques of mass spectrometry

1921 Construction of a national road grid

1925 Mass-produced diesel-powered
agricultural tractor

1928 Electron microscope

1933 FM broadcasting (a method of radio
broadcasting using frequency modulation)

1935 Rural Electrification Administration
(REA)

1945 Food and Agriculture Organization (FAO)
of the United Nations

1954 Silicon solar cell

1957 Sputnik 1

1961 Integrated circuits (used in Apollo missions)

1962	Active, direct-relay communications, commercial satellite
1972	Earth Resources Technology Satellite (Landsat)
1974	Glyphosate first sold
1977	*Voyager* spacecraft
1989	Modern GPS satellite
1990	High-speed T1 transatlantic internet connection (between Cornell University and CERN)
1990	First web browser, called WorldWideWeb
1991	Li-ion battery
1991	Linux operating system
1993	Internet Engineering Task Force (IETF), an open standards nonprofit organization
1995	Modern image chip (CMOS sensor with PPD technology)
1999	SD card
2001	Wikipedia and MediaWiki tool kit
2003	Human Genome Project completion
2012	First-generation open-hardware computer (Raspberry Pi Model B) running Linux
2014	#DR Pixhawk drone autopilot
2022	FarmOS recognition as a digital public good by the United Nations

used together, our observation, analysis, and communications tools begin to illuminate our world, just like a friend shining a flashlight in a dense wood can help us find a path. We each share the same light and can then share insights. Imagine trying to bake your first batch of sourdough bread without the benefit of a recipe? Then imagine a recipe for the care and feeding of an acre of rich biological diversity, both above and below ground. The "yeast" in this case is the billions of organisms in the soil, plants, and air.

Adaptation Rather than Prescription

Many people believe that humans have insufficient knowledge to manage natural systems. They often reference large-scale geoengineering projects such as seeding oceans with iron or the upper atmosphere with particles to manipulate the greenhouse effect. These are massive and risky endeavors that quite rightly should be questioned. However, it is also true that humans have been massively manipulating the environment for tens of thousands of years, often without clear intention, which has led to mass plant and animal extinctions and changing weather and water patterns across the globe. Fortunately, there are many counterexamples of people stewarding thriving ecosystems for millennia. In the developed world, most of us have become divorced from direct management of nature, including food production. At the same time, our impact on the natural world continues to grow dramatically. Our separation from nature has hidden our impact from our eyes, giving the false impression that ecological systems are in good shape. But it has also obscured our ability to understand which practices and tools have worked and which ones have failed.

A practice or a tool can fail for many reasons (including personal ones), but perhaps the most common reason is the lack of proper context—a land's soils, water patterns, and vegetation, for example. Context adds complexity. A simple practice or tool might be preferred—a plow, for instance—but if it's the wrong tool for the context, or it is used in the wrong way, the result can be disastrous. There's a long history of using prescriptive "recipes" for farming and ranching as pushed by agricultural specialists, including the USDA, land grant colleges, consulting experts, and industry. However, in many cases the recipes were the wrong ones or were not detailed enough to be adapted to the nuances of local conditions. This was a critique of the initial FAO approach to knowledge-sharing. The increased use of pesticides and synthetic fertilizers, for instance, has been a staple of many modern agricultural prescriptions. Alternative prescriptions, even if they use some of the same ingredients and tools, are often met with resistance even if the benefits are clear and provable. We are at the point, however, where we must find a way to create sources of information that can be trusted and are outside the control of the usual gatekeepers of knowledge.

The question is not *can* we manage nature through agriculture and commerce but *how* we manage nature and commerce through our agricultural tools to achieve the best outcomes. The solution cannot be no governance but must be good governance; not no management, but good management. The knowledge utility is a recognition of a special connecting role that some tools have in making whole systems function. Even the most ardent free-market advocates might grudgingly acknowledge the crucial role of the stock exchange registry as a utility supporting trusted transactions. A competitive football game is only a football game because of the shared adherence to the rules by which it is played. Likewise, stable climate, healthy wildlife populations, productive soils, and clean air and water are shared, if often neglected, utilities.

The history of agriculture and knowledge utilities also has precedent in the founding of the USDA. Henry Leavitt Ellsworth, a commissioner of patents in 1837, became interested in improving agriculture. He began by collecting new varieties of seeds and plants and sending them to members of Congress and agricultural societies. Later, he established the Agricultural Division within the US Patent and Trademark Office (at the time, under the Department of State) and implemented the collection of agricultural statistics. Eventually, he called for a public depository to preserve and distribute the new seeds and plants as well as for the preparation of statewide reports about crops in different regions. His efforts paid off in 1862 when President Lincoln signed legislation establishing the USDA, which he called the "People's Department." Three years later, the USDA built its first headquarters, near the Mall in Washington, DC, and subsequently rolled out an outreach program that included cooperative extension services and land grant universities across the nation.

A cornerstone in the original building is still visible in the current USDA headquarters. It is engraved with the motto "Agriculture is at the foundation of Manufacture and Commerce." In launching these universities and extension programs and using this language, the USDA recognized the key economic and strategic value of agricultural knowledge exchange. In the context of a nation being torn apart by extractive agriculture of another kind, in the form of slavery, the words take on additional meaning. While not fully realized at the time, or perhaps the department was not visionary enough, the motto remained on letterhead until the early 1980s when the utility function of agriculture was papered over and replaced with the phi-losophy of agriculture *as* commerce. Agriculture—and by that word I mean the social, economic, and ecological utility of food production—was for a time considered (and behaved) just like any another industry. But the unpaid balance from that debt has now come due.

I realized that the prescriptive "recipe" approach to farming, which has been applied to not only conventional agriculture but organic as well, was inadequate for the complex tasks required of the Great Regeneration. Any ingredient list of tools and practices for a farm or ranch by definition would be too short and too rigid and therefore ultimately inadequate for the job. What we need are more collaborative languages and tools to describe our changing world and respond to the new opportunities emerging as a result of fast-moving developments in technology, economics, and environmental change. Advances in soil microbiology, for example, are opening up vast new horizons below the ground's surface. However, many land management

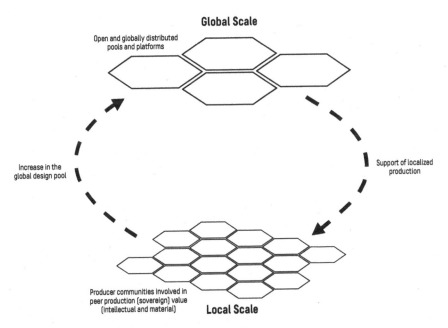

A Circulatory System of Global Knowledge for Local Production. An open-design commons enables design exchange for everything, from medicine to machines, to solve real problems. Shared global knowledge for local production is a virtuous cycle. *Adapted from Jose Ramos.*

recipes, even progressive ones, focus solely on what is happening at the soil surface and above—plants, animals, insects, rain, heat, snow, floods, and wildfire. We now know that the biological universe underground is just as critical, if not more so, to the health and productivity of the land, just as a new understanding of our own microbiome is moving us beyond a prescriptive approach to human health to tailor treatments to an individual's own genetics and flora. A good recipe captures the essence of a complex process while also allowing it to be adapted to local ingredients and preferences.

The patterns we are creating form an intergenerational commonwealth of agricultural knowledge—a never-ending story that can be handed down, containing our best understanding of how we interact with each other and our environment. A commonwealth is something we can create only together, and it requires intention and energy to be maintained. Unlike the restricted section of a research library, the new agricultural knowledge commons is being coded into a new kind of public square and invites all to view and contribute to the patterns we are each most drawn to. It is created by each of our individual contributions that influence those around us, as well as contributing to the larger pattern of the whole, which can be rewritten, overwritten, and ever expanded.

A single spade of soil is home to billions of living organisms. *Tuckaway Farm.*

Millions of years of genetic source code adapted to local environments can be held in your hand and can easily be stored, gifted, and exchanged.

Living roots of grasses and legumes together gather abundant carbon, water, and nitrogen from the atmosphere to grow soil.

Microscopic patterns of life are reflected in landscape patterns of life—here, in rice cultivation. *Planet Labs.*

Patterns of life are repeated in our human representations at different scales, not unlike this Byzantine tree of life mosaic from the sixth century. *Carole Raddato.*

Hexagonal patterns of carbon-based life on the landscape are visible in ancient cities such as Palmanova in eastern Italy. *Ulderica Da Pozzo.*

Wolfe's Neck Center for Agriculture and the Environment creates a carbon foot-path, an experiential learning landscape that spans human management of natural systems across freshwater, saltwater, field and forest, urban and rural, local and global communities. *Wolfe's Neck Center.*

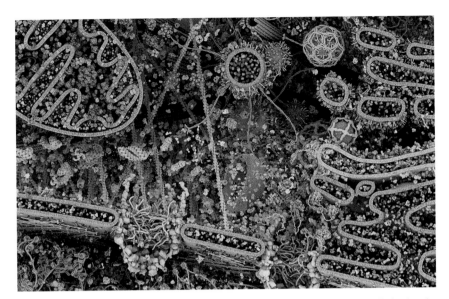

We now have the technology to observe the patterns of life at the sub-cellular level, as well as at the regional and planetary levels, and then make connections, inferences, and breakthroughs that were not previously possible. *Cell Signaling Technology.*

The patterns ranging from the microscopic to the global start in the field. Leah Puro uses an open-source, handheld soil spectrometer at Wolfe's Neck Center. *Wolfe's Neck Center.*

A Public Lab balloon is prepared to carry a camera over research plots, enabling us to elevate our perspective as part of a new public observatory.

Using a simple weather balloon it is possible to elevate our perspective to 90,000 feet, where the thin blue edge of the atmosphere is visible across the curvature of the Earth's horizon. *Overlook Horizon.*

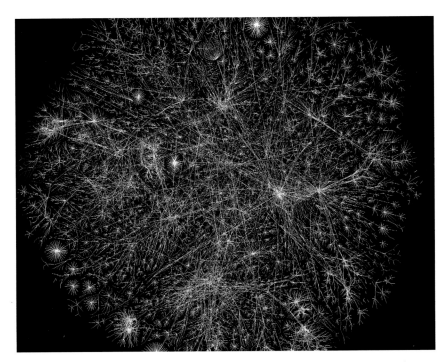

Silicon structures combine to form patterns of the internet that mirror complex biological structures created over just the last 4,000 days. *The OPTE Project.*

The move from theory to practice is expressed, regardless of technology, in the action we take to mix our labor with the millions of years of genetic history that are then expressed in the patterns of living roots in the land we steward.

CHAPTER 5

The Art and Science of Collaboration

Does a civilization fall when the soil fails to produce, or does a soil fail only when the people living on it no longer know how to manage their civilization?

—CHARLES KELLOGG, *Soils and Men: Yearbook of Agriculture* (1938)

The key to successful regeneration is to out collaborate the competition.
—GREGORY LANDUA, CEO, Regen Network

In 2014, three farms located near my own family farm in New Hampshire decided to work together to deliver products to the same restaurants, thereby increasing their collective capacity to provide local food while saving time, fuel, and money. They called themselves the Three River Farmers Alliance, and initially they worked from the back of a single delivery truck. The idea caught on, and the farms built a simple ordering application so that chefs, schools, and hospitals could see the current availability of produce and place orders, while each farm still maintained its individual identity and control of pricing, even as they shared ordering, billing, and delivery logistics. Demand grew quickly, and the alliance added new farms and local food producers. By 2020, forty farms were working collaboratively through the alliance to deliver food year-round to more than 100 restaurants, schools, grocery stores, and other wholesale buyers. The original, single delivery vehicle became five refrigerated trucks. Twenty people were hired as staff to handle the increased volume.

The alliance operates as a multifarm CSA (community-supported agriculture) as well, delivering weekly shares of seasonal produce to individuals and families in the area. A shared online store lists an extensive selection of products from member farms, including meats, cheeses, sauces, pasta, and fermented foods, adjusted on a weekly basis according to availability. The complexity was made manageable through shared communications and logistics software. All products, packing lists, printed labels, delivery routes, and invoices are seamlessly managed through software that results in digitally purchased products being bundled and transported to customers while also making it clear which product came from which farm. Additionally, sales can be scaled while maintaining a sense of origin and relationship with the producer through a shared communications platform that documents what is available, from whom, where, and at what price. Some of the efficiency of Enterprise Management Systems (EMS) once reserved for Fortune 500 companies was making a difference for a few small farmers directly servicing local markets. Orders grew, and the geographic radius of the alliance expanded to include northeastern Massachusetts and the southernmost part of Maine.

Then COVID-19 hit in March 2020. Restricted to take-out orders, restaurants cut their food orders dramatically. Some shut their doors for good. Schools suspended their purchases. Other institutions followed. Suddenly, the business model that the alliance had nurtured into existence was turned on its head. Then an unusual thing happened—people began to worry about their own food supplies. Products disappeared from grocery store shelves. Staple foods, including potatoes, grains, and eggs, were in short supply. Phones began ringing at alliance-member farms, and new customers began appearing at local farm stands looking for food to buy. At the same time, stay-at-home orders and social distancing measures meant many people were unable or unwilling to venture very far for their sustenance. It raised questions about long supply chains, COVID-19 outbreaks at industrial food-processing sites, and community resilience in the face of a surging pandemic.

In response to the crisis, the Three River Farmers Alliance decided to adapt its digital platform and transportation model to accommodate home delivery, bringing food directly to consumers. This new system shifted restaurants into a new role as a value-added production stop. New distribution routes were created within a week. COVID-19 safety became part of the mission. "No-touch" boxes of food were now dropped off at homes or delivered to collection points. New orders poured in, as did memberships: from a few hundred accounts to thousands in a few weeks. "People want safe food," one farmer observed. Local food, in other words. Today, the alliance's farm produce is delivered to food banks, pantries, and regional hubs. Farm stands have become mini farmers markets. The foundation of their ability to handle the complexity of local markets can be found in their tech-enabled collaboration and their use of newly available knowledge utilities.

This collaborative business model built on the development of shared software serves as an example of what is possible within an expansive future of resilient and regenerative local food systems and the accompanying benefits that come with fresh food grown in healthy soil by farms who are interconnected with customers both digitally and physically. It also opens up opportunities for connection, not just as related to food distribution but also for how communities function and support what is important in times of plenty and in times of crisis. Out of adversity comes opportunity, especially for ideas that have stood the test of time, such as local food. The Three River Farmers Alliance is a fascinating case study of a marketplace becoming a community platform where customers could not only purchase food and communicate directly with farms but also contribute and support the food system directly. Options were added online so members could contribute extra money to help defer transportation costs for people on limited incomes or who had suddenly lost their jobs. Elderly people and others at risk of contracting COVID-19 had access to fresh food, delivered to them directly. Restaurants made meal kits with local food products so people could prepare their favorite meals at home. As the COVID-19 pandemic wanes, the cross-pollination and replication of aspects of this model are evolving into autonomous distributed networks that connect to one another at various scales and fill new niches as they specialize, overlap, and intersect.

Cooperatives

Cooperatives are an ancient form of collective activity, but they can now be deployed at all scales and in increasingly innovative ways. In this chapter we will look at examples of contemporary collaboration made possible by knowledge utilities and the long history of cooperatives and collaborative technology in agriculture. Cooperative social structures themselves are also important collaborative tools for facilitating exchange.

Markets, which are indeed cooperative social structures with agreed-upon rules, can be competitive or collaborative depending on their goals. Additionally, many of the lessons gleaned from the development of knowledge utilities can be adapted to serve the goals of market creation while increasing the transparency of information about place, quality, price, origin, and production practices. At their core, marketplaces are an exchange of knowledge in a standardized structure involving a process of asking questions, finding answers, and then making decisions. Markets are about data, context, and levels of trust reflected in the level of accounting and in the form of currency debt, which itself is a collaborative construct to facilitate exchange.

It is not surprising that software apps that are used to help develop local agricultural markets are using some of the same knowledge utility functions and foundations. The process of sharing an environmental observation with a neighbor (such as a date when cover crops started blooming) can use the same knowledge utility underpinnings that enable the sharing of a similar observation with a research university, which also wields the same tools that can be used to inform a market decision for when seed harvest is likely as well as the projected availability of a crop.

Today, we are seeing a resurgence of organizing around food system cooperatives, from Cincinnati to Detroit to other cities across the United States and around the globe. Co-ops are unique forms of solidarity that have been an essential part of creating democratic systems in the US. This effort blossomed in the 1970s as member-owned food cooperatives began providing access to food that was largely excluded from grocery stores, including organic products, whole grains, and macrobiotics. Co-ops also challenged the industrial food system with educational programs about the risks of pesticides and genetically modified organisms. Co-ops were early champions of local food and fair labor practices. Currently, there are 148 organizational members of the National Cooperative Grocers Association, representing 1.3 million individual members and combined annual sales of nearly $2 billion.

Co-ops are not just for groceries. Local and regional cooperatives include worker-owned manufacturing operations, depositor-owned credit unions, agricultural marketing and purchasing co-ops (such as Organic Valley), and rural electric co-ops, which provide 40 percent of the nation's electric distribution.

In 1985, hoping to forge a closer relationship with their customers, two farmers in western Massachusetts opened one of the first CSA programs. Today, there are nearly 9,000 farmers markets across the nation and over 7,000 CSAs.

The first successful cooperative of any sort was organized in 1844 in Rochdale, England, when a group of weavers and craftsmen pushed back against the tide of industrialism sweeping the nation by opening a store to collectively sell their products. They called themselves the Rochdale Society of Equitable Pioneers, and they authored a set of principles that have recently been updated by the International Cooperative Alliance. They include:

- open and voluntary membership
- democratic control
- economic participation by members
- autonomy and independence
- education, information, and training
- cooperation among cooperatives
- concern for community

Today, cooperatives are all around us, often hiding in plain sight. Overall, there are nearly 30,000 cooperatives in the United States, accounting for two million jobs and $500 billion in annual revenues. IRS-recognized categories include consumer cooperatives, which are owned by the people who buy their products or use their services (REI is the nation's largest example); producer cooperatives, set up so that farmers and others can sell their products under one label (Organic Valley, for example); purchasing cooperatives, for businesses working together in order to be competitive with national chains (like the members of the National Cooperative Grocers Association); and worker cooperatives, which are owned and run by their employees.

The consumer cooperative category is by far the largest in the United States, and the movement as a whole is gaining momentum. Recent research suggests why. The broad and diverse benefits created by co-ops make them resilient in a crisis. Credit unions, for example, survived the Great Recession of 2008 relatively unscathed because they viewed mortgage speculation as contrary to the interests of their members. Consumer cooperatives mostly focus on essentials necessary to a healthy society: food, water, electricity, insurance, and finance. Their primary mission is to provide public services, not to act as engines for wealth accumulation. This public-service orientation is why it is not such a big leap to extend the cooperative model to ecological restoration and carbon sequestration.

Largely forgotten in these stories are the contributions of Indigenous people, who have a rich tradition of cooperative governance that might be considered an advanced cultural tool kit. One very good example is the coordination shown by the people of the Longhouse and the people of the Iroquois Confederacy. Members of the Iroquois Confederacy and Huron inspired our own US Constitution and all three branches of government and influenced the conversations of prerevolutionary French salons and the works of Enlightenment thinkers such as Jean-Jacques Rousseau and Benjamin Franklin.[1]

Mutual aid expresses itself as a form of collaborative technology in unexpected ways in our history. For example, the Underground Railroad of the nineteenth century, which enabled enslaved African Americans to escape north into Canada, created a shared and cooperative network of safe houses. After emancipation this network formed the foundation of African American mutual aid, resulting in the founding of benevolent societies to develop businesses, buy land for farms, and create schools. W. E. B. Du Bois, George Washington Carver, and Ida B. Wells addressed these issues in their day, just as cooperatives for Black farmers are doing today. Civil rights leaders such as John Lewis, Ella Baker, and Fannie Lou Hamer honed their organizing skills within cooperatives.

We now have the opportunity to build on that shared American experience. The organizations that helped usher in rural electrification, such as the Grange and Farmers Union, can inform the idea and practice of modern knowledge utilities. We have digital tools that can be designed for cooperation, and they can exhibit the strength that comes from participation and equity, rather than isolation, fear, segregation, and competition. Democratizing these tools means they can enhance rather than diminish tech-enabled regenerative agriculture and bring it to a global scale.

Agroecology: A Trip to Rome!

Knowledge utilities are possible in many forms and scales. They can be managed through tradition and ritual or through new means such as digital services. Agroecology is one such knowledge utility that is bridging traditional and digital knowledge. Agroecology is a food and fiber production system that is based on the ability of nature to regenerate and grow abundance if the proper niches are filled in the ecosystem. It stands in contrast to industrial farming's modus operandi of mining and depleting scarce resources, such as soil and nutrients. Agroecology models food systems on nature, starting with the vast abundance of sunlight and the power of photosynthesis and microbial processes to harness and transform solar energy into nutritious food that sustains us. Agroecological systems, which often include trees and perennial plants, are intentional, intensive, and integrated, often mixing different types of crops together in one plot or field. To the maximum extent possible they mimic nature, including diverse plant and animal species, mosaics of landscape types, year-round production, groundcover protection to suppress weeds, natural composting of leaf litter and other organic material, and the creation of suitable habitat for beneficial insects.

Agroecology is not new. It has been employed by peoples around the planet for millennia to produce food, clothes, timber, medicine, fodder, and fuel. Indigenous practices continue to provide these products today in many places and serve as both inspiration and teaching models for modern agroecological endeavors. This is not simply a suite of regenerative practices, however. Agroecology is as much about families, communities, social bonds, cultural traditions, spiritual beliefs, and economic resilience as it is about specific aspects of farming or ranching.

In the context of a knowledge utility, we can explore the potential of agroecology in terms of guiding principles for a new kind of public science, built not on the proprietary allocation of scarce resources (mined or manufactured), but on a collaborative system focused on growing abundance through knowledge and working within natural limits. Agroecological principles underlie regeneration and the next advancements of agriculture. While there are many ways of approaching the beneficial intersection of people and nature—such as permaculture, holistic management, sustainability, resilience, and now regenerative agriculture—I find that the foundation of agroecology as a science is one that best bridges people and their environment. As a framework, agroecology defines both the questions we ask and the tools we use to answer them.

This hit home in 2019 when I had the privilege of attending the second International Symposium on Agroecology for Food Security and Nutrition at the headquarters of the UN's Food and Agriculture Organization (FAO) in Rome. It was a remarkable juxtaposition to have a meeting over coffee on a roof deck overlooking the ruins of the imperial Roman chariot track that is just a short walk from the Colosseum and the site of the Roman Senate. I was struck by how current conversations about agroecology, including global access to land and the struggles between farmers and merchants for fair prices and fair treatment, echoed similar conversations that took place more than 2,000 years earlier in the same place, though now without mob violence!

At that time 2,000 years ago, a contentious legislative process led by Tiberius Gracchus broke down over a proposed agricultural land reform law, known as Lex Agraria, which aimed to redistribute farms and large landholdings of Roman merchants. As the empire grew, the merchants had vastly expanded the size of their estates using slave labor shipped in from conquered territories and accumulated land previously held by regular citizens and yeoman farmer-soldiers. As retired soldiers began returning home from tours at far-flung outposts across the empire, they found their farms in poor repair due to long absences, or they discovered their land lost altogether to the newly rich merchants. The legislative breakdown around the popular land reform and proposed redistribution led to a violent merchant-organized mob that stormed the senate. The riotous insurrectionists broke off the legs of chairs and benches to use as makeshift clubs (weapons were not allowed in the Senate chamber) and beat to death Gracchus and fifty of his fellow reformers. This led to the end of land reform and the beginning of the end of the Roman Empire.

It was a moving and humbling experience to participate in the FAO conversation to identify key proposals for scaling up agroecology to meet not just climate challenges but also wider development goals such as social well-being and health, all while literally standing on the ruins of a fallen empire, just blocks from where Lex Agraria was violently ended. The process of bringing forward recommendations for action was based on the outcomes of regional processes involving the more than 600 participants speaking dozens of languages.

The FAO's hosting of the dialogue on agroecology was symbolic of a broad willingness by people around the world to look beyond industrial solutions to our problems. The chairman of the meeting urged the FAO to "take the lead, in partnership with other international organizations, academia, and research organizations, to facilitate the development of new methodologies and indicators to measure . . . agricultural and food systems beyond yield at landscape or farm level, based on the 10 Elements of Agroecology . . . and work in a coordinated way to scale up agroecology through policies, science, investment, technical support, and awareness."[2] He also called for an effective and massive transfer of knowledge to millions of farmers, while creating networks for family farmers to share their innovations and also creating multi-stakeholder platforms for collaboration at local, national, regional, and global levels.

The Ten Elements of Agroecology are principles that help define a place-based, systems-level approach to agriculture that promotes the continual improvement, conservation, and restoration of ecosystem services, including carbon and water storage, nutrient cycling, biodiversity, resistance to pests and disease, animal welfare, and social justice and equity. Each principle has continual improvement goals that can be measured. However, to measure and improve each of the indicators behind the principles will require a new circulatory system for ideas, information, and inspiration.

The FAO's Ten Elements of Agroecology are:

1. **DIVERSITY:** Agroecological systems optimize the diversity of species and genetic resources, including intercropping, crop-livestock systems that rely on local breeds adapted to specific environments, and, in aquatic environments, traditional fish polyculture farming.

2. **CO-CREATION AND SHARING OF KNOWLEDGE:** Agroecology depends on context-specific knowledge. Its practices are tailored to fit the environmental, social, economic, cultural, and political context. Sharing of knowledge plays a central role in developing and implementing innovations. Through the co-creation process, agroecology blends traditional and Indigenous knowledge, producers' and traders' practical knowledge, and global scientific knowledge.

3. **SYNERGIES:** By optimizing biological synergies, agroecological practices enhance ecological functions, leading to greater resource-use efficiency

and resilience. For example, in Asia integrated rice systems combine rice cultivation with other products such as fish, ducks, and trees. By maximizing synergies, integrated rice systems significantly improve yields, dietary diversity, weed control, soil structure, and fertility, as well as provide a biodiverse habitat and pest control.

4. **EFFICIENCY:** Agroecological systems improve the use of natural resources, especially those that are abundant and free, such as solar radiation, atmospheric carbon, and nitrogen. By enhancing biological processes and recycling biomass, nutrients, and water, producers are able to use fewer external resources, thus reducing costs and the negative environmental impacts of their use.

5. **RECYCLING:** Waste is a human concept—it does not exist in natural ecosystems. By imitating natural ecosystems, agroecological practices support the biological processes that drive the recycling of nutrients, biomass, and water, thereby increasing resource efficiency and minimizing waste and pollution. Recycling can take place at both farm-scale and within landscapes through diversification and the building of synergies between different components and activities.

6. **RESILIENCE:** Diversified agroecological systems have a greater capacity to recover from disturbances, including extreme weather events such as drought, floods, and hurricanes, and to resist pests and diseases. Following Hurricane Mitch in Central America in 1998, biodiverse farms implementing agroforestry, contour farming, and cover cropping suffered less erosion and retained 20 to 40 percent more topsoil than neighboring farms practicing conventional monocultures.

7. **HUMAN AND SOCIAL VALUES:** Agroecology places a strong emphasis on human and social values, such as dignity, equity, inclusion, and justice. It empowers people and communities to overcome poverty, hunger, and malnutrition while promoting human rights, such as the right to food. Agroecology seeks to address gender inequalities by creating opportunities for women, who play a vital role in household food security, dietary diversity, and health as well as the conservation and sustainable use of biological diversity.

8. **CULTURE AND FOOD TRADITIONS:** Culture and food traditions play a central role in society. However, our current food systems have created a disconnection between food habits and culture. This disconnection

has contributed to a situation where hunger and obesity exist side by side in a world that produces enough food to feed its entire population. Almost 800 million people worldwide are chronically hungry, and 2 billion suffer micronutrient deficiencies. By addressing the imbalances in our food systems, agroecology can move us toward a zero-hunger world.

9. **RESPONSIBLE GOVERNANCE:** Transparent, accountable, and inclusive governance mechanisms are necessary to create an enabling environment that supports producers to transform their systems following agroecological concepts and practices. Successful examples include school meals and public procurement programs, market regulations allowing for branding of differentiated produce, and subsidies and incentives for ecosystem services.

10. **CIRCULAR AND SOLIDARITY ECONOMY:** Strengthening short food circuits can increase the incomes of food producers while maintaining a fair price for consumers. These include new innovative business models alongside the more traditional territorial markets where most smallholders market their products. Social and institutional innovations play a key role in encouraging agroecological production and consumption. Examples of innovations that help link producers and consumers include participatory guarantee schemes, local producer's markets, denomination of origin labeling, community-supported agriculture, and e-commerce schemes.

As exciting as it was to feel part of a larger movement, even more important was the sense of hope I felt in the broad recognition of solutions coming from the care and curation of natural systems scaled up to a planetary level as an evolution beyond industrialism. I saw that a more sophisticated and nuanced approach to agroecology principles and supporting science could be adapted to local experiences and conditions. The meeting in Rome highlighted a shift in emphasis from the control of scarce resources to a framing of governance and distribution of abundance. In this shift, I saw very clearly the role that open-source knowledge utilities could play in bringing together observational tools, analysis, communications, and technology to act as a foundation for providing democratized knowledge for local production. I saw governance in action as representatives of industrialized nations and

Indigenous civil society groups worked together to move a unifying concept forward. In my mind, the 600 people in attendance represented a microcosm of the governance and convening power that would enable the democratization of every watershed. Like nature, we can build on pathways we know work already. Nature, and now people, have a legacy of abundant knowledge we can draw upon in the form of genetic diversity and open-source code as a utility upon which we can innovate—and so, too, can we build on governance patterns that work.

Here was an effort that both implicitly and explicitly laid out the case for a new kind of knowledge utility based on the principles of agroecological design that would be necessary to fully realize the goals of regeneration and grow our commonwealths of nature and of knowledge. It built on the last centralized attempts the FAO started over sixty years prior to create new extension-based knowledge pathways, and it called out new forms of knowledge-sharing and collaboration as a key innovation that would also require new tools and ways of using those tools to collect, communicate, and exchange knowledge.

The Advent of OpenTEAM

Soon after the meeting in Rome, I was able to see these agroecological principles put into practice in innovative ways, building on ancient forms of agricultural cooperatives. I saw this process of pre- and post-competitive structure play out directly as a project leader for the Open Technology Ecosystem for Agricultural Management (OpenTEAM). Launched in 2019, OpenTEAM stewards a new ecosystem of critical knowledge utilities. This collaborative community of skilled professionals was founded by Wolfe's Neck Center, LandPKS/USDA Agricultural Research Service, and Stonyfield Organic, with matching funding through the Foundation for Food and Agricultural Research. OpenTEAM brings together people and organizations who might otherwise be considered competitors but who nonetheless have convened around key questions that are better answered together. OpenTEAM builds on the foundation of the open-source movement and existing networks and communities that formed the internet as well as more recent communities, such as Farm Hack, farmOS, the Gathering for Open Agricultural Technology (GOAT), Purdue Open Ag Technology

Systems Center, the AgStack Foundation, and the Linux Foundation. Each of these communities, of course, builds on thousands upon thousands of other open-source efforts.

The OpenTEAM community has grown to represent a global network of more than forty-five organizations, including research universities, food companies, environmental markets, agricultural foundations, agricultural technology companies, and agricultural community organizations. They are unified by a vision of agriculture knowledge commons as a shared, multigenerational human endeavor and public science. It has grown internationally, representing diverse productions systems, scales, cultures, and geographies. From the initial twelve members, OpenTEAM grew rapidly to include not just progressive organic farmer organizations such as Pasa Sustainable Agriculture in Pennsylvania and Quivira Coalition in the southwestern US, but also large food companies such as General Mills, research universities, government agencies, ag-tech companies, and newly forming environmental-service market players. The goal of OpenTEAM is to leverage existing capacity across the members' organizations and enable the seamless exchange of agricultural ideas, information, and inspiration across geographies, production systems, and cultural boundaries.

The OpenTEAM collaborative was designed to work not just at the technological level but to prototype an approach to working at the global, systemic level, as articulated in the FAO agroecology symposium in Rome just a few years prior. In those few short years, OpenTEAM grew from just an idea to a global collaborative, partnering and influencing agricultural software projects of governments as well as other private and public enterprises around the world. We have seen how strengthening the roots of an agricultural knowledge commons is finally making the supporting infrastructure visible to the community. In chapter 7 we will explore examples of a common core of software and services that make up what might be called a digital equity tool kit. The success of the OpenTEAM example has helped provide the stability, will, and inspiration required to support rapid exploration and innovation. OpenTEAM is just one example of many similar decentralized social, technological, and economic systems that will be necessary to scale the commonwealth of agricultural knowledge globally and transition to the agroecology-led growth and regeneration we now know is possible.

As these new ways of organizing and analyzing emerge, we have the opportunity to take a fresh look at the resources we can manage for abundance. Many of the technologies and collaboration strategies that OpenTEAM uses were developed in the search for the elements and life on other planets. So it is fitting that we also train those same tools on our own planet for a fresh look at the abundant resources we have to work with. As we see the dividends from sharing abundant knowledge and innovation accumulate, we can think again about abundance, scarcity, and the health of life on Earth.

CHAPTER 6

The Elements of Abundance

Whoever could make two ears of corn, or two blades of grass, to grow upon a spot of ground where only one grew before, would deserve better of mankind, and do more essential service to his country, than the whole race of politicians put together.

—JONATHAN SWIFT, *Gulliver's Travels* (1726)

There is perhaps no better demonstration of the folly of human conceits than this distant image of our tiny world. To me, it underscores our responsibility to deal more kindly with one another, and to preserve and cherish the pale blue dot, the only home we've ever known.

—CARL SAGAN, *Pale Blue Dot: A Vision of the Human Future in Space* (1994)

On my family farm in New Hampshire, we put up thousands of bales of dry hay every year. Raking hay gives me lots of time to think as I drive the open tractor in concentric patterns around the field. One of the things that I think about is the power of the sun, the stored fossil power in the diesel fuel running the tractor, and the photosynthesis energy that hay represents that might be turned into beef or milk and eventually my own food and energy later that year.

The dominant economic theory, in place since the industrial revolution, says we live in a world of scarcity. We are told there isn't enough oil, food, or rare minerals to go around, and so the free market, often boosted by government subsidies, research, or legislation, decides who gets what scarce resource. The truth is we live in a world of abundance, not scarcity—an abundance of sunlight, soil, microbiota, carbon, human ingenuity, and many other resources. In Economics 101 we encounter the simplified idea that scarcity affects the price of a resource, often driving it upward if the supply of a resource is limited or can't meet demand. The reality, however, is far more interesting and complex.

Some resources, of course, *are* scarce. Gold, for example, is perhaps the best-known rare mineral. It's employed in mobile phones, laptops, batteries, electric vehicles, and many other products, including medicine. Gold can't be made synthetically; it must be mined in nature, but once melted down it remains stable as a consistent, testable commodity that is fungible. An ounce of gold is the same material anywhere on Earth and can be tested as such. However the perception of a resource's scarcity, and the ability to exchange it for value somewhere far from where the resource was produced, is often more important for concentrating power than the resource's actual abundance or intrinsic value. There are other reasons for perceived or short-term scarcity, of course. Famines are frequently not caused by actual food shortages but by political, social, or administrative bottlenecks—if not as a means of war and violence in a less obvious form.

Nitrogen is a good example of how scarcity theory works in agriculture, much to our detriment, which we will cover in more detail later in this chapter. In the atmosphere, nitrogen is a highly abundant gas, comprising almost 80 percent of the air we breathe. It is also an essential nutrient for plants, required for their growth and vigor. Most plants, including nearly all commercial crops, cannot access nitrogen directly from the air. Instead, nature came up with an ingenious way of getting the nutrient to plants through their roots, using fungi in the soil that make nitrogen available in exchange for plants' carbon-based sugars. These fungi acquire nitrogen molecules from a class of bacteria that gathers the nutrient from the roots of certain plants, shrubs, and trees that can absorb the gas from the atmosphere through their leaves. It is an elegant natural symbiosis among microbes, fungi, and plants—and a key to regenerative agriculture. In contrast, the absence of nitrogen-fixing plants in conventional agriculture, such as in mono-cropped fields of corn, means the nutrient must be applied as an artificial fertilizer, produced with fossil fuels. Thus, a false scarcity is created by our actions. Nitrogen is abundant, but we choose to make it scarce—to the profit of industry. At the same time we create a massive pollution problem when the fertilizer washes downstream to create dead zones.

It bears repeating: The truth is, we live in a world of abundance, not scarcity. We only have to change our context, our point of view, our perspective to see it. This change means changing incentives, and recognizing scarcity that is generated by the political economy versus real scarcity. Recognizing the difference between the two can also change how we treat one another and how we approach the environment. Abundance by itself does not imply optimal distribution of resources or automatically lead to better decisions. Indeed, food, land, and water scarcity, as experienced at the local level, are real, but it is important for everyone on Earth to understand what is truly scarce and what is artificially scarce. Our commonwealth of knowledge provides the context that true scarcity exists just above our horizon as the edge of the biosphere transitions into the depths of lifeless space. This transformation in how we think about abundance and scarcity begins with our relationship to the soil below our feet, followed by our relationships to the plants, people, and animals that sustain our healthy ecosystems.

Health and Abundance

Let's start by reflecting on health. How do we define health? There are physical indicators linked to technical medical definitions of health, but we know it is much more than that. I like to think of health as a state of well-being. It involves emotional, social, economic, and other aspects of a healthy life, not just blood tests and EKGs. Health is a positive condition and not solely defined by the lack of illness. And it is cumulative—many things contribute to our well-being and they build on each other. Which raises a question: Is there an upper limit to well-being? No, but only if we shift our thinking about health as a scarce resource to one focused on abundance. What might that look like, exactly, in people or nature? Scientific evidence now supports what so many have suspected for so long: human health and the health of our environment are tightly linked. It is not surprising that *health* as a concept and as a word is now being applied more widely to natural systems. It is also not surprising that we are still struggling to adapt our economic systems to concepts of well-being. For example, the United States continues to have a hard time matching the concept of "consumerism" to health. Too often, our consumer culture and economy actively work against our well-being.

We now know that photosynthesis, growing roots, and feeding microbes are the key building blocks of living soils and of soil health. Soil health is the ability of soil to function biologically, chemically, and physically to sustain productivity and environmental quality as well as promote plant, animal, and human health. These key activities are part of the original conditions for an oxygenated planet and the foundation for a stable climate that supports abundant life. That abundance is often obscured in surprising ways. For example, if agricultural commodities, such as wheat or corn, have a bumper year resulting in a surplus, the result is generally a drop in price for the farmer. Abundance, in this model, is a form of penalty. However, if health is the goal rather than commodity sales, the situation changes. There is always demand for greater health. The creation of a "surplus" of photosynthesis or soil carbon offers nearly unlimited opportunities for meaningful work for people anywhere on the planet.

In this chapter, we focus on what is abundant. In addition to economics, we need to transition from a "conservation" movement that protects nature from people (the scarcity view) to a place of regeneration where people and nature work together (abundance). This involves a radical evolution from a Western industrial perspective, but such an evolution is not so radical when we take a longer view of human history and look to the relationships of Indigenous and migratory people with the land. Much of what has been claimed to have been "wilderness" has in reality been stewarded and managed by people for thousands of years. For example, the abundance and bounty that was observed by European settlers in the Americas was not an accident but the deliberate management of hunting and fishing grounds and agricultural lands. In 1997, during a visit to a farm in the pampas outside Buenos Aires, I saw flocks of birds so thick they would block the sun, and the noise was such that it was necessary to shout to carry on a conversation. It was the first time I could imagine the natural abundance described by European colonists: salmon so thick you could cross a river by walking on their backs.

Counter to what we are taught in Economics 101, the question is much less about the distribution of *scarce* resources and more about the distribution of *abundant* resources, with an emphasis on cycles and equity. The focus on regeneration does not contradict observations regarding inherent limits to growth and the alarming level of global extraction that has taken place, but rather provides an evolutionary pathway for growing a complementary and cyclical system not based on mining and pollution. This shift, I believe, has powerful ripple effects. The scarcity perspective focuses on lowering the cost of extraction and concentrating the benefit (profit), while an abundance perspective looks at the problems and the questions we ask more as challenges of distribution. As economist Elinor Ostrom demonstrated in her work with collaboratives, abundance fights the power of extraction and promotes collaborative governance of common resources instead.

To move to a world focused on producing abundance and health, the concept of our commonwealth of agricultural knowledge becomes a key resource from which to draw. I think of land stewardship as working with an artist's palette, drawing from our commonwealth of carbon, nitrogen, phosphorus, water, and biodiversity to mix into various shades of green, blue, and brown. The big swaths of color created in the form of life on land and in oceans are punctuated with brilliant displays of life, from the crimson flowers of clover and the deep purple of vetch to the white of peas and yellow of mustard and sunflower. Like life itself, there are infinite ways to combine these abundant colors, which each represent biodiversity and stewardship of the land, but they also share common hues and patterns. They exist in roughly similar proportions, and yet each land steward's interpretation of them, and each season, creates unique expressions of color. Here, let's examine the abundant elements that create the colors of life all around us. We will start with power, followed by water, carbon, nitrogen, oxygen, phosphorus, biodiversity and genetic code, and land.

Power

Enough solar energy reaches the Earth in one hour to fill all the human economy's current energy needs for a full year.[1] This observation is the foundation of all else, and so we will continually come back to it to illustrate that the challenge is not scarcity but rather concentration and distribution. The inequitable distribution of mined and scarce resources, such as fossil fuels, is well known. Ironically, coal, oil, and gas are the products of abundance. They are concentrated forms of ancient sunlight captured through millions of years by plants via photosynthesis and stored in carbon bands and oil deposits in relatively few places around the world. Solar power is both abundant and, for the most part, its core colors (photons) are non-rival, meaning they are free in equal amounts to anyone and anything clever enough to harness them. With the combination of photosynthesis and photovoltaics, we are currently only harnessing a minute amount of the abundant power available to us. It even seems that the areas where power may be needed most to restore the landscapes are generally also those with abundant solar power. We are seeing this phenomenon now. A crude parallel is how the need for air conditioning to cool buildings is elegantly matched in areas with an abundance of sunlight. So, we will say it again, the sun's energy reaching Earth in any given hour is sufficient to power the entire human economy for a whole year. Power is not scarce, but access to the means to harness power and distribute it is not yet ubiquitous.

Embodied energy is defined as the energy consumed by all of the processes associated with the production of an object, from the mining and processing of natural resources to manufacturing, transport, and delivery of the final product. However, much of the critique of embedded energy in products comes from calculations based on nonrenewable sources. While there are certainly valid critiques of embedded oil and gas power used to create products, the emphasis might rather be on how we allocate clean, renewable energy and the utility of the products themselves. The concern, if we think more broadly, might be less about the embedded pollution in the creation of the iPhone, for example, and more about the utility of that phone over its lifetime. Does communications infrastructure like a phone substitute for transportation, the purchase of a car, or office heating costs, for example? Are we using solar power generation to support ground water extraction or cryptocurrency mining (Bitcoin consumes half as much energy as all the world's data centers at the moment, which is roughly 0.6 percent of global electricity consumption[2]), or are we using it to create landscapes, governance, communications, and utilities that have a positive feedback loop and increase the velocity of recycling? The total amount of power used is important, but how it is generated and how it is used are even more important.

By thinking of our available technology not as a substitute but as a complement to natural systems, we can begin to design systems that do not perpetuate the current extraction-based economy. If we focus only on the technology and not the context, we may continue to accelerate global degradation, only with solar-powered industrial extraction. Historically, many civilizations created massive soil degradation using organic farming methods (by over-tilling with the plow, principally). Success depends on how the technology is used. An electric bulldozer run on renewable energy may beneficially terraform a water catchment or destructively fill a wetland. An electric chainsaw may harvest wood for biochar production or clear a rainforest. Low-cost solar water pumps are now being used to deplete groundwater reservoirs in many developing countries.

When the focus is on extraction, energy is generally cited as the limiting factor—in other words, energy to access minerals, energy to concentrate and refine—so the economics of mining is often stated in terms of the cost in power. Oil sands and deep-ocean drilling require more power to access oil than was needed to tap the oil seeping out of the ground in Pennsylvania in the 1800s. Most rock and seawater have minute concentrations of valuable minerals. At different times attempts have been made to refine gold, lithium, phosphorus, and other rare and valuable minerals from seawater; however, the energy cost to retrieve the minerals was perceived as being higher than the value of the minerals themselves. We also know that we are using only a fraction of a fraction of a fraction of the energy available to us, and only using a fraction of a fraction of a fraction of our available solutions. This is why we have an opportunity to change the framing of abundance and energy.

This opportunity leads us to ask new questions: How might we measure the total horsepower of the sun available on a given acre? What portion of that acre is green with plants and what portion covered by solar panels? What percentage of that captured energy then goes to regenerating the system?

Power is the rate of work, measured as work per unit of time. In the 1770s, James Watt, a Scottish inventor who improved the steam engine for industrial use, found that a brewery horse could lift 180 pounds (by turning a shaft) 180 feet in one minute. This became known as horsepower. It is calculated today at about 746 watts (the unit is named for the inventor, James Watt) or 32,400 foot-pounds per minute. We are most familiar with horsepower in terms of tractors, lawn mowers, and other vehicles. For the sake of understanding abundance, the United Nations Development Programme, in its World Energy Assessment in 2000, found the annual potential of solar energy to be as much as 49,837 exajoules (EJ)—many orders of magnitude more than the industrial economy uses.

We can translate exajoules into a more familiar term of horsepower. At any average moment, about 1,850 horsepower of solar energy reaches the atmosphere above an average acre. Sunlight energy is not mechanical energy but arrives as electromagnetic radiation, mostly light of various wavelengths, a portion of which is absorbed by the atmosphere. About a third of this radiation does no work but is merely reflected back into space by clouds, snow, and ice; some warms our atmosphere, and the remaining power reaches the surfaces of the Earth—ocean, rock, soil, a blade of grass, your cheek. If all the remaining solar energy could be converted to mechanical power by the time sunlight reaches the Earth's surface, it would represent about 1,300 horsepower per acre. At that rate, if fully captured, it would take only about four square feet to power a typical US home. Those four square feet are about .02 percent of a typical suburban home's footprint. Certainly, we will not capture 100 percent of the energy on every square foot of the planet's surface. However, the illustration makes the point that energy and the distribution of energy is not scarce, only the knowledge to harvest it.

We could interpret green on the Earth's surface as an indication of a portion of energy being transformed into life with the regenerative power of the sun's fusion reactor. We have abundant pathways to capture more of this abundant power, but let's focus on a few methods that mix the ancient with the new. We can think about layering photosynthesis with photovoltaics in complementary patterns with the intention to both capture sunlight and translate it into usable energy. French researchers have pioneered a method to successfully combine food and solar energy production on a single plot of land. Photovoltaic panels are not just placed side by side with crops, but above them as well. Agricultural space is at a premium in densely populated Europe, so researchers came up with a way to place panels above crops on firm metal stands so as to maximize the amount of electricity produced by the panels while minimizing the decline in plant growth caused by the shade. The panels can also tilt vertically to let in more sunlight when needed. The goal is to take advantage of the massive abundance of solar radiation available every day, both as a source of renewable energy and to produce food to eat.

With an agroecology perspective we can grow healthy landscapes with an understanding of how solar installations can be designed to reduce surface temperatures, form wind breaks to enhance microclimates for growing plants, increase water capture, create edges for biodiversity, and harness photosynthesis. In Georgia, White Oak Pastures, one of the best-known regenerative farms in the United States, has partnered with an enterprise called Silicon Ranch, which is now one of the largest mobile sheep flocks in the state and was formed specifically to graze under solar panels. Grazing and managing productive grasslands is proving, not surprisingly, to be more economical for the solar companies than mowing or herbicides. More exciting still is to see biological and silicon energy collection layering harmoniously on the same landscape.

Water

Of all the water on the planet, only 3 percent is freshwater and only 1 percent is drinkable, the rest being locked up in glaciers, ice, and deep aquifers. That figure might make it sound like water is a scarce resource (and in a desert it is), but there is more than enough freshwater for all humans, plants, and wildlife, even without taking into consideration how much is squandered by poor management and lost to evaporation every year. Most of the water we

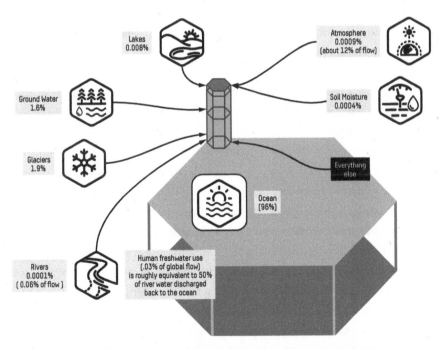

Global Water in Different Forms and Abundance. The great global abundance of water points to the importance of managing the flow and productive storage of water across our landscape to support life.

drink and use comes from rivers or is pulled from shallow aquifers, but there is water in the soil below our feet and in the air all around us. Using agroecological practices and regeneration, we can increase the amount of fresh water available—and even change the water cycle to increase local water availability.

The cooling effect of trees can be experienced by anyone enjoying refreshing shade, or tested by simply using a kitchen thermometer in tilled ground and then in soil under even a small amount of mulch to measure the drastically reduced temperature under the mulch. On my own farm, I placed small data loggers on ground cultivated for asparagus and then just a few feet away under a bit of straw mulch for strawberries. During the day, the bare soil around the asparagus was 120°F; the temperature under the mulch was just 74°F. It was instantly clear to me which environment was likely to retain more water and be more conducive to life.

While the water cycle is itself a biogeochemical cycle, the flow of water over the Earth is a key component of the cycling of other biogeochemicals such as nitrogen, carbon, and phosphorus from land to water bodies. The numbers on water flow through the landscape are staggering. There is 3,000 times more precipitation hitting the soil than is retained in soil moisture, which contains more than 7 times as much water as humans currently cycle annually. Amazingly, there is 4 times more water retained as soil moisture than in rivers, and 1.4 times more water reaches our soils than evaporates. On land, 70 percent of water evaporates and is cycled back into the atmosphere.

The point here is that over 2,000 times more water evaporates into the air than is retained in the soil, and so small changes in surface temperatures and soil moisture retention can make a big difference. Larger amounts of water cycling through leaves and roots, more moisture condensing near the ground, and more surface water in catchment basins can have large effects in regulating climate and available water for photosynthesis. Of course, management approaches will vary substantially in accordance with local conditions. Areas currently experiencing three to eight inches of rain a year would, at first glance, have little in common with those that get forty inches or more per year. However, many strategies for retaining water in soil and determining how it moves through the landscape are independent of the volume of precipitation. Water, soil, aggregates, and roots actually follow the same principles in tropical rainforest as in arid rangeland, which means the colors on a palette may be similar, but the local adaptation may have many different shades.

Living plants modify and slow water cycling in many ways, particularly on land. Plants, along with affiliate communities below and above ground, create and maintain water-holding fertile soils and reduce surface temperature, and therefore act as both a sponge and an umbrella. Surface vegetative debris, called *litter*, in combination with biologically produced sticky macromolecules, holds soil-aggregate building blocks together, which in turn slow the huge, potentially destructive force of moving water. This has ripple effects, from slowing and moderating stream flow across regions to reducing bank erosion downstream as well as preventing sedimentation and nutrient loading at vast river deltas. The cooling effect of plant transpiration also modifies water cycles at the surface and near-surface levels, which keeps temperatures within optimal ranges of many biologic processes. We are all keenly aware of the narrow temperature range of our own body's comfort zone, which is tightly tied to our bacterial microbiome's comfort zone, and we feel its effects when we have a fever or when ambient temperatures fluctuate by more than 10°F.

Substituting gas or diesel water pumps with solar pumps without addressing the fundamentals of the water cycle across the landscape does not result in abundance. Instead, it illustrates a political drought created by the disconnect of governance and nature. There is sufficient water in the world, but it is a matter of governance and rights in terms of how it is stored, moved, used, and reused. There are examples across the United States and globally of water governance systems that have worked for generations, while other arrangements have led to conflict and drought. In this, as in so many areas, the issue is not the resource itself but how it is managed. The solutions that work tend to be products of good governance.

Water vapor is a potent greenhouse gas, but it is also part of a global transport system that distributes moisture to us, and that system can be influenced by how we manage plants and soils. You can test this in your own backyard with an infrared thermometer. Simply point the thermometer at bare ground, or at soil under thick grass, and note the reading; then dig down eight inches and point the thermometer again. The difference will be dramatic, similar to what I found with the straw mulch on my family's farm. The soil under cover and deeper down is dramatically cooler than the exposed soil.

Efforts to manage water is as old as civilization and not unique to humans, as anyone who has seen a beaver dam can attest to. We have access to both ancient and modern tools to create artificial aquifers underground that do not evaporate but rather cool soil and buildings. Ancient Persian architecture and Roman aquifers creatively used the topography of the land to move and store water. The USDA's core soil health principles published by the Soil Health Division, which echo Gabe Brown's principles, apply here as well:[3] Living roots, diverse cover, and reduced disturbance all are important tools in water management. Increasing soil health is like creating a lush living sponge that can soak up water to mitigate the destructive effect of large rain events, while reducing leaching, erosion, and runoff.

In 1995, John Liu, noted Chinese-American filmmaker and ecologist, filmed an area in the Loess Plateau near the city of Xi'an in central China to document its transformation from a barren and eroded landscape into an oasis through a combination of land forming and planting to affect the local water cycle. It was at that point Liu noticed the possibility of humans restoring ecosystems, rather than only destroying them. Liu became an ecosystem ambassador for the Commonland Foundation based in the Netherlands. In 2017 he founded Ecosystem Restoration Camps, a worldwide movement that aims to restore damaged ecosystems on a large scale. Terraforming is often discussed at the planetary scale by advocates of Mars settlements, and yet we likely first came to understand the concept as children, creating dams and rivulets in small streams with sticks and stones. Beavers, with their dams, transform entire landscapes to their benefit at a scale only matched by humans. Like so many patterns of the water cycle, it's a matter of scale. Principles that we use to transform our backyard with trees and raised beds can also apply to managing a million-acre ranch. In both cases we are creating combinations of living soil, roots, and cover to help slow the movement of water across the surface so it has time to infiltrate, recharge, and not run off or leach, which can then scale to the level of a whole watershed. We can each work and steward our scale of management within a mosaic that creates larger repeating and recognizable patterns.

Carbon

Carbon in all its forms is the structural lattice for life on Earth, from the lignin in tree trunks and the complex blood vessels in animals to the carbon dioxide

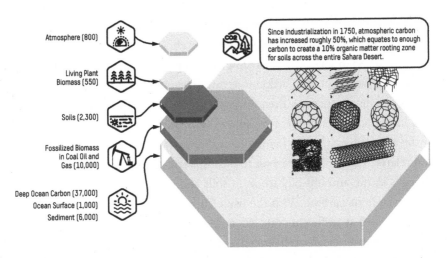

Atmosphere (800)

Living Plant
Biomass (550)

Soils (2,300)

Fossilized Biomass
in Coal Oil and
Gas (10,000)

Deep Ocean Carbon (37,000)
Ocean Surface (1,000)
Sediment (6,000)

Since industrialization in 1750, atmospheric carbon
has increased roughly 50%, which equates to enough
carbon to create a 10% organic matter rooting zone
for soils across the entire Sahara Desert.

Global Carbon in Different Forms and Abundance (in Gigatons). Almost three times more carbon is in soil than in the atmosphere, and more than four times more carbon is in soils than stored in living plants. That means even small changes in soil carbon can have large effects on atmospheric carbon.

abundant in the atmosphere, plants, soils, and oceans. As we look at the pools of carbon, they each represent vastly more than is being used even in abundant life forms. It is remarkable how carbon in some forms is considered scarce, such as diamonds or oil, but considered a surplus when it comes to atmospheric carbon. The question, as with so many of our abundant building blocks of life, is where and in what form can carbon be most productive for regeneration. Increasingly, we are beginning to understand the methods needed for transforming forms of carbon, even to the point of the largest diamond company announcing that it will no longer sell mined diamonds but rather will manufacture them by transforming other forms of carbon.

Atmospheric carbon, which is the essential fuel for all life forms, is transformed by the process of photosynthesis. This process splits water and forms carbohydrates, using carbon dioxide from the atmosphere. These carbohydrates and their derivatives are constantly being oxidized by many forms of digestion, decay, and combustion that sooner or later returns CO_2 to the atmosphere.

Even the ancient carbon pools now locked up in oil and gas, rather than being transformed into heat and emissions, might also be turned into tools for moving nutrients and water. For example, plastics used for greenhouse coverings can be used to create regenerative microclimates. Similarly, we can use these carbon pools to fuel equipment for terraforming the landscape with terraces and ponds to store more water. If we choose to use those reserves of ancient carbon in service of balancing carbon pools, rather than just further extracting, we might then yield more carbon capture than emissions.

For reference, there is already roughly more than three times the carbon in the soil than in the atmosphere, and over 50 percent more carbon in the atmosphere than in living plants. We now know that concentrations and relative ratios of soil carbon to atmospheric carbon are such that small changes in the amount of carbon in soils and plants can have large effects on atmospheric carbon. With the use of non-carbon-based energy sources, when utilized at sufficient scale, those large effects can erase the notion of carbon scarcity (usually referenced in terms of "national oil reserves"). In terms of accessing the basic building blocks of life, carbon can be considered an ubiquitous surplus in the atmosphere and so provides a clear pathway to address deficits in soil and in biodiversity, both places where it can be expressed most productively. Of course, when we hear about the surplus of atmospheric carbon, we think of it as a liability rather than as a resource to cycle back to where it is needed—namely in the soil or in life forms. It is these ubiquitous carbon bonds that create cellular structures and lignin that enable a sequoia to root deep in the soil to stabilize itself against winds while also towering above the surface of the Earth. The cycle of carbon through the various possible forms of life is constant in our daily living experience as we eat and breathe and our bodies repair themselves. Yet it is only through the regenerative agriculture lens that we can see the importance of building soils to further increase the carbon sponge of organic matter, humus, and in living tissues ranging from kelp to roots, leaves, and branches.

Nitrogen

Nitrogen is the most common pure element on Earth, making up nearly 80 percent of the entire volume of the atmosphere. We all balance our carbon-to-nitrogen ratio in our carbohydrate and protein diets, which also happen to be the ratios of the diets our own microbiome prefers. In our own diet, we are most familiar with nitrogen in the form of high-protein foods such as beans and meat. And yet nitrogen scarcity in agriculture led to some of the most consequential developments of the twentieth century with the invention of synthetic nitrogen fertilizers. Despite its abundance in the atmosphere, nitrogen is not very abundant in the Earth's crust. When mineral nitrate became scarce during World War I, the industrial synthesis of it became common for the manufacture of explosives and fertilizer. Nitrogen compounds are synonymous with proteins and constantly interchange between the atmosphere and living organisms; however, nitrogen must first be "fixed" into a plant-usable form. Mostly this is done by diazotrophic bacteria through enzymes known as nitrogenases. When the ammonia is taken up by plants, it is used to synthesize proteins. These plants are then digested by animals that use the nitrogen compounds to synthesize their own proteins and excrete nitrogen-bearing waste, which is consumed by other lifeforms. Finally, billions of soil microorganisms die and decompose, undergoing bacterial and environmental oxidation and denitrification, returning free nitrogen and the other elements of life back to the atmosphere.

Excess nitrogen-bearing waste, when leached, leads to eutrophication and the creation of marine dead zones because nitrogen-driven bacterial growth depletes water oxygen to the point that all higher organisms die. Furthermore, nitrous oxide, which is produced, for example, when soils are deprived of oxygen from flooding, results in denitrification and potent greenhouse gas emissions. While nitrogen is abundant, the limiting factor to availability is in improving the rate of "fixing" and capturing of nitrogen by plant roots and of transforming nitrogen into an exchangeable form as plant and animal proteins. Not surprisingly, the solution lies with a more active nitrogen-fixing microbiome.

By observing the unique patterns and colors that these elements of life create, we can also begin to see when we are successful at enhancing the development of a functioning microbiome. On my own farm I can see these elements coming together above ground when the deep red of a crimson clover blossoms as part of a cover crop. I can dig below the surface of the soil and see the elements of life in the white nodules of nitrogen-fixing bacteria and in the darker brown and black spots accumulating around thread roots deep in the soil profile, where atmospheric carbon has been transferred from plant to microbe before being transformed into soil.

The Haber-Bosch process of manufacturing nitrogen fertilizer takes roughly thirty million BTUs per ton of nitrogen created, as in urea or ammonia (the energy equivalent to around 200 gallons of diesel, though the fuel here is usually in the form of mined natural gas). In contrast, soil microbes are fed atmospheric carbon captured in photosynthesis powered by solar energy. The obvious appeal of industrial nitrogen is that it can be transported and applied where it is needed without requiring the complexity of the microbiome and all the needs of soil life and health. However, bypassing the complexity has its costs. Nitrogen fertilizer applied in surplus acts like fast food and makes other biological systems lazy. When we are on road trips, fast food is useful in the short term to satiate hunger, but such foods are not part of a long-term strategy for health. By focusing on healthy soils, we can create stable aggregates, built on sticky proteins, to provide nitrogen-fixing bacteria the space they need to thrive and function. Healthy soils will also create animal and human systems that cycle nitrogen locally so that concentrations of our own nutrient streams from cities and farms do not accumulate in our drinking water sources and pollute them but find their way back productively to growing plants.

Oxygen

Oxygen is the second most common component of the Earth's atmosphere, after nitrogen. Earth is unusual among the planets of our solar system in having such a high concentration of oxygen gas in its atmosphere. Under normal circumstances oxygen is freely available to us, but we are keenly aware of each breath when we consider how susceptible we are to having it taken away by disease or other means. Although we all have universal access to oxygen gas, we forget that soil organisms also breathe and require oxygen to thrive. Oxygen is drawn into the soil through capillary action as water infiltrates and moves through the profile, where the oxygen is stored or leached out. This action is crucial to growing the diverse communities of microbial life that mine and build soil aggregates. These microbes also need to eat and drink, and it's partly why soil disturbance that creates compaction—destroying pathways and spaces—is so detrimental.

Oxygen is continuously replenished in Earth's atmosphere by photosynthesis, which uses the energy of sunlight to produce oxygen from water and carbon dioxide. Oxygen is too chemically reactive to remain a free element in air without being continuously replenished by the photosynthetic action of living organisms. Most of the mass of living organisms is oxygen as a component of water, which itself is the major constituent of lifeforms. It seems we yet again have a situation where abundance is tied to enhancing the productivity of plants and roots in healthy soils.

Phosphorus

Phosphorus is the eleventh most abundant mineral on Earth. Phosphorus, unlike the other primary building blocks of life (carbon, water, nitrogen, and oxygen), is not atmospherically available. However, because it is primarily transported in water and by moving around feed, fuel, and animals, many of the soil-health approaches related to the water cycle and input reduction, such as regenerative grazing, also work to improve phosphorus-use efficiency. In the Earth's crust, phosphorus is widely distributed in many minerals, with the highest concentrations in phosphate rock, which is today the primary commercial source of this element as a fertilizer. Large phosphate deposits remain in the world, more than three hundred billion tons by some estimates. According to the US Geological Survey, about 85 percent of Earth's known reserves of phosphate rock are in Morocco, with smaller deposits in China, Russia, the United States, and elsewhere. Assuming no increase in the rate of use, the reserves would last for 260 years.

Some have put forth the concept of "peak" phosphorus, suggesting that we may have already reached the maximum global extraction rate of phosphorus as an industrial and commercial raw material. Current extraction and use rates effectively illustrate the potential limitations of mining from any stockpile. They also highlight the limitations of an economic system that ships the agricultural products globally in such a way that nutrients cannot circulate and instead result in scarce mineral concentrations, not as a valuable resource but as a waste product and environmental liability. The algae blooms in the Gulf of Mexico at the mouth of the Mississippi river illustrate this phenomenon. However, that same analysis also builds the case for cycling and recycling what is still abundant and also for transitioning to methods that move phosphorus within the environment with better water, manure, and urine cycling. History contains many examples of cities cycling phosphorus and nitrogen in urine and manure into the countryside. Perhaps most notable is the complex nutrient management programs in London during World War II. This effort was documented in the 1954 book *Reconstruction by Way of the Soil* by G. T. Wrench.

The warnings of "peak" phosphorus do not take into account the biological, regenerative, and agroecological approach. Various physiological strategies are used by plants and associated microorganisms for obtaining phosphorus from even low concentration levels. We now know that it is possible to improve a plant's ability to "mine" phosphorus in relationship with biologically active fungal and bacterial associations in the soil. The most abundant primary source of phosphorus in the crust is the mineral apatite, which can be dissolved by natural acids generated by soil microbes and fungi. The dissolved phosphorus is bioavailable to terrestrial organisms and plants, and then returns to the soil after their decay in a form more available to other organisms. Phosphorus retention by soil minerals is usually viewed as the most important processes in controlling phosphorus bioavailability in the soil. With very few tweaks to how phosphorus is provided to plants (which in turn provide it to animals), there is no reason to think that phosphorus will become a limit to global regeneration.

Let's look at a few ways in which phosphorus can be captured, cycled, and recycled.

ORGANIC SOURCES

Urine, bone ash, and guano were historically important sources of phosphorus. Urine contains high concentrations of plant-available phosphorus, a quality that is still harnessed today as fertilizer in some places. However, manure and urine from livestock can pose a pollution problem when they're concentrated in feedlots, where feed is transported in from afar. By moving animals closer to where the feed is grown, much, if not all, phosphorus needs can be met locally. If animals are able to graze, then they transport, cycle, and apply the phosphorus right where it is needed—without inputs.

Humans are a big part of the phosphorus cycle as well, and so our own nutrient management also comes into play. The oldest method of recycling phosphorus is through the reuse of animal manure and human excreta in agriculture. In this method, phosphorus in the food is excreted, and the animal or human wastes are subsequently collected and reapplied to the fields. Although this method has maintained civilizations for centuries, the current system of manure management is not logistically geared toward application to crop fields on a large scale. At present, manure application could not meet the phosphorus needs of large-scale agriculture. Nevertheless, it is still an efficient method of recycling phosphorus and returning it to the soil and is an important biological primer in developing regenerative systems.

BIOLOGICALLY ACTIVE SOIL

Locked-up phosphorus is not available to the plant unless the right organisms, such as mycorrhizal fungi, are present to mine or mineralize the nutrient for the plant. Biologically active soil provides the diverse microbial ecosystem at the roots to create a complex economy of exchange of mined minerals for other resources they need, such as carbon.

AQUACULTURE AND ALGAE CYCLING

Phosphorus gets pooled naturally in aquatic systems. As regenerative principles are applied to land-based agriculture, they are also being expanded and explored in aquatic-based regenerative aquaculture. Algae can be harvested and converted into feed and supplements, providing a way to cycle phosphorus efficiently. Kelp and other large brown-algae seaweeds, for example, have a cell structure that filters seawater, enriching the oceans with nutrients, including nitrogen, potassium, and phosphorus. Due to this constant filtration, kelp plants grow at very fast rates, as much as half a meter a day. This rapid growth rate makes kelp a renewable and ample resource for not only many sea creatures but also as an organic fertilizer and a source of over 70 vitamins and minerals. With two-thirds of all sunlight landing on the Earth's oceans and seawater containing concentrations of both macro- and micronutrients to support life, it makes sense to use the abundant life of the oceans to harness sunlight through photosynthesis to filter and concentrate these nutrients. We currently harvest and remove vast quantities of minerals into our terrestrial system in the form of seaweed and seafood, and we have done so for thousands of years. For example, in North America, Native Americans today are still cultivating varieties of flint corn that were traditionally fertilized using fish and seaweed.

IMPROVED APPLICATION AND SOIL HEALTH

Mined phosphorus must be made into a soluble form to be available to plants. In general, as little as 20 percent of phosphorus applied is taken up by plants in the first year, and the excess contributes to water pollution downstream. The best solution is cycling and storing phosphorus in living cycles on land

and ocean where it is most available and using plants as well as bacterial and fungal processes to capture phosphorus in new microwater cycles rather than from shipping rock all over the world.

Biodiversity and Genetic Code

Just as photosynthesis is necessary to transform solar energy into living organisms, so biodiversity is necessary for life to adapt to local conditions, generate healthy soils, and drive water, carbon, nitrogen, phosphorus cycles. Biodiversity is built on the basic building blocks coded in DNA, which itself is made up of hydrogen, oxygen, carbon, nitrogen, and phosphorus. Other than phosphorus, all these core elements are ubiquitously available in the atmosphere. Phosphorus, as covered previously, is also abundant and highly recyclable.

All the elements of life form the raw materials made available and useful to the land steward through our commonwealth of nature and knowledge. When I choose a hard, red winter wheat that I know will thrive in our wet, humid New England summers and yield well, I am drawing on millions of years of genetic code made up of these five elements, which was itself selected by hundreds of years of selective breeding across continents and generations and was refined in the last ten years by friends and neighbors across the region. This pattern of adaptation I participated in is not unique but happens every year everywhere.

There is biodiversity that has locally adapted to create life from photosynthesis in nearly every environment on Earth. Every seed in every environment carries the result of millions of years of versions of that genetic code that can be germinated and expressed above and below ground. But unlike a silicon SD memory card, that seed also contains the mechanisms and energy to build and run that code to remanufacture itself. The richness of this code is still being identified, of course, and although we have diminished our repositories of genetic code through extinctions, biodiversity remains a key and abundant resource. And this genetic diversity of code can be transported. When I save and share the seed of Narragansett flint corn, I am also sharing in the thousands of growing seasons that were managed by the people who lived on the land before me. These seeds can be used and reused to regenerate soils. This is a key piece of the living elements of life available in a land stewardship palette. As we germinate seeds we have shared, we are translating the code in each seed into a human-readable format in the colors and shapes of the leaves, seeds, flowers, and roots. We can now share the expression, and our observations, not only in more seed but also digitally in the forms of plant data services and new knowledge utilities to even more fully value and share this abundance and diversity.

Genetic diversity is key to forming a living carbon sponge and stable soil aggregates for bacteria and fungi, for concentrating minerals, and for creating more local carbon and water cycles. Nature cycles nutrients in tight loops and has a positive feedback effect. It's like a bank putting more currency into local circulation, which further stimulates the economy. Biodiversity is not just something that happened but is something that is happening as an ongoing process. Despite the threat of mass extinction brought on by the Great Acceleration, biodiversity—and our ability to exchange knowledge about available biodiversity—is still abundant and can be made more so. Just twenty-eight commodity crops, with sugar cane and beet, wheat, rice, corn / maize, and soybean at the top, contribute 90 percent of US food plant supplies. It is amazing to think that we are using just twenty-eight species out of the approximately 320,000 known species of edible plants to feed ourselves and restore our landscapes, and that the number of plant species is dwarfed by the diversity of microbes and insects. And yet all that diversity carries the same elemental building blocks that form the palette of abundant, diverse life available to each land steward.

Land

Farmers have only temporary control over their land. It can be theirs for a lifetime and no longer. The public's interest, however, goes on and on, endlessly, if nations are to endure.

—HUGH HAMMOND BENNETT (1959)[4]

In a global narrative of rising real estate prices, mass migration, and refugee crises, our first instinct may be to think of land as the ultimate scarce resource. Isn't land scarce, particularly agricultural land? Surely land must be the limiting factor to regeneration? We aren't making new land after all, right? But maybe we are! It depends on your definition of land. Is it merely a measurement of surface area? Or is it the amount of area available for cultivation? What if we enlarged our definition to include trees, shrubs, grass blades, and animals? What about the soil underfoot? Isn't that land as well? Isn't land really best evaluated in three dimensions? Healthy soil can be shallow or deep, depending on soil type and vegetation native to the area. How might we consider the surface of land in tidal areas that has been managed for aquaculture for centuries? What about aquaculture and the management and cultivation of the water about the ocean floor? These questions each stretch our ideas of boundaries.

If previously degraded land can be restored to health—and thereby increase the amount of carbon it holds—then we might say the amount of land available to us for stewardship can be expanded, potentially by a great deal. Instead of being viewed through a scarcity lens, degraded land becomes something else altogether when we look at it through a perspective of abundance, as we identify opportunities to increase photosynthesis and store carbon productively. Land available for repair and regeneration has potential, in other words, to be abundant. Necessary minerals may be in place already. Plants and trees and biodiversity are available. All the oxygen, nitrogen, and carbon dioxide that is required exists for free—and in abundance—in the air.

If we look at the land that we actually occupy and the land that is actively managed—the tiny part of our commonwealth of nature that we steward—then this resource looks far less scarce. Again, this is not a new idea. At the same moment the *Encyclopédie* was being published, a group of natural philosophers in France called physiocrats began forming a theory that all wealth came from the soil and all other economic activity was a transformation of that original wealth. The key tenet was expressed by the Marquis de Mirabeau who wrote, "Land is the source or material from which Wealth is extracted . . . human labour is the form which produces it; and Wealth in itself is no other than the sustenance, the conveniences, and the comforts of life." The term *physiocrat*, which derives from the union of the Greek words *physis* (nature) and *kratos* (power), was coined to emphasize the basic thesis of this school of thought: It assigned productive power to nature, or to the fertility of the land, which through agriculture created greater wealth than what was expended to produce it. Land was not fixed but *variable* and could be improved through stewardship.

There is, in fact, no real upper limit to soil carbon. Amazonian *terra preta* soils, the ancient precursor to biochar, were packed with carbon (as much as 80 percent carbon content!) and ran meters deep. Are we so humble in our

Land Area Available for Regeneration, 2015 (excluding ocean-based aquaculture)

	Land Cover (in hectares)	Percentage
Tree-covered areas	4,336	29.11%
Grassland	3,478	23.35%
Terrestrial barren land	1,884	12.65%
Herbaceous crops	1,713	11.50%
Shrub-covered areas	1,628	10.93%
Natural sparsely vegetated areas	868	5.83%
Inland water bodies	444	2.98%
Woody crops	200	1.34%
Shrubs and herbaceous vegetation, aquatic or regularly flooded	185	1.24%
Permanent snow and glaciers	85	0.57%
Artificial surfaces (including urban and associated areas)	55	0.37%
Mangroves	19	0.13%
Total Land Area	**14,895**	**100%**

aspirations that we cannot conceive of accomplishing what South American civilizations accomplished thousands of years ago and without the benefit of a global agricultural knowledge commons? I often hear skeptics discuss the issue of soil carbon saturation, and I respond by saying we should be so lucky to run out of land to saturate with carbon. This is precisely the kind of problem we would benefit from having. Besides, what are the downsides in trying?

I have experienced an evolution in my own thinking about land management, scarcity, and abundance. Formerly, in managing our own hay fields, I would have a hard time not thinking short-term about how many cuttings I might be able to get, worrying about the weather and about the price that could reduce the crops' profitability. However, my thinking changed when I signed a contract with the USDA to create a functional ecosystem that would help protect the quality of drinking water and establish habitat for pollinators and threatened species. My thinking shifted from a mindset of what land was available that might be brought into production for grain and hay to a mindset of what degraded land at the edge of the fields might be available to be improved. This was a fundamental shift for me in viewing the landscape, and this flipped the script for where I saw economic opportunity. While on a tractor raking hay one day, I started looking increasingly beyond the field edges and broadened my thinking of degraded land and began to view it as being abundant.

When we include the millions of acres of degraded land in our understanding of our combined global resource, then the scope of a cross-generational effort emerges. It is an effort with an infinite level of creativity and detail for meaningful work for generations to explore and refine so as to improve the known upper limits of soil health and the health of the ecosystem.

In the scarcity mindset, some have suggested that intensified use of land for agriculture could free up other areas to be "rewilded" back to nature. This approach abdicates our responsibility for restoring the natural function of land onto the majority of our degraded landscapes. Much of what is considered "wilderness" has in fact been managed by people for millennia. Even the concept of "wilderness" separates humans from nature and denies us the opportunity to act beneficially. It reinforces the belief that humans will always be extractive, parasitic organisms. Wilderness implies that land must be tamed, rather than signaling the continuity of life that is ubiquitous in nature. Wilderness implies otherness. If we expand our view of land as being manageable through stewardship—for foraging, for cultivation, for restoration, and for increasing the commonwealth of nature—then the resources we have available shift from a sense of scarcity to an abundance of opportunity, each acre being a new chance to improve the diversity and productivity of life on Earth.

CHAPTER 7

The Technology of Trust

In managing the commonwealth of nature, our big problem is that we tend to treat the truly scarce as if it were non-scarce. The opposite problem arises with the commonwealth of knowledge, in which we tend to treat what is truly not scarce as if it were.

—HERMAN DALY

One of my most rewarding discoveries in the early days of Farm Hack and Public Lab was the realization that I was not alone in enjoying the process of sharing work with others! Sharing work, in this context, subverts power centralization. It is a choice to create managed competition of ideas and then focus the powerful marketplace profit motive on the skilled implementation of ideas to create profit from practice, not protection. In my observation, the exciting enterprises are focused on profit from production and innovation rather than protection of intellectual property. Certainly, the early architects of our now-outdated global patent system had similar worries. Our commonwealth of agricultural knowledge is a byproduct of managing many knowledge utilities together to enlarge the trusted knowledge held in common. Universal access to trusted knowledge also serves as a necessary tool for large-scale collaboration and is a necessary precondition for groups to share ideas and ask more questions effectively. Groups like Public Lab and Farm Hack thrive when ideas become abundant and there are social rewards for sharing. It is a bit like the camp-song refrain that love is like a magic penny—if you give it away, it comes back with more.

Practically, it means balancing the process of accessing and contributing to the global knowledge commons while maintaining local control and sovereignty for ourselves, our watersheds, and our bioregions. It means stewarding trust to speed up our learning process through sharing peer learning, and it means benchmarking our efforts so that we may first learn locally in order to learn globally, and then to discern the repeated, recognizable patterns of the commonwealth of nature. This also means using some of our extracted resources to build, maintain, and improve infrastructure in order to help us each stack time, as well as experience the benefit of the millions of lifetimes that make up our commonwealth of agricultural knowledge. This chapter will look at the social, technical, and physical infrastructure required for this commonwealth to exist and grow.

In 2008, I was in the midst of designing a complex, modular, self-contained biodiesel processor based on designs I had seen in Argentina. I found that the challenge was not to create an equipment company but to rapidly problem-solve and create tools and processes that worked. I was faced with a choice in the design process. The first tendency, dominant at the time, was to develop in isolation, then protect and monetize the idea by treating it like a scarce resource. This would be done through legal methods such as patenting and licensing. The second approach was to share the idea widely and invite collaboration and critique, treating ideas as abundant and not in need of protection but rather as improving with exposure and use. I chose the latter pathway. By shifting into thinking of ideas as abundant rather than scarce, it was possible to benefit from the creative and social process by which shared ideas spread, appreciate, and improve. This is not to say that all knowledge should be free and shared without controls—quite the contrary. However, when hard problems, questions, and solutions are shared, then we can move from a transaction economy based on scarcity to a relationship economy based on gifting, gratitude, and the imprecise social accounting of the reciprocity of small favors.

When we are working on something too big to tackle alone, precise transactions and compensation becomes less important, less transactional, more relational, and in my experience more enjoyable and rewarding. Complex reciprocity, a term used in open-source development systems, is a description of an alternative set of powerful incentives wherein contributions that community members make to the commonwealth are rewarded, but often indirectly. The reciprocity, however, is that the rewards are meaningful and enriching enough to build trust and enable communities to be constructed around the principle. Communities such as OpenTEAM and Farm Hack thrive on complex reciprocity. The question is, how do we carry that sense of trust and shared purpose across organizations and communities that we have not yet met, or may never meet, in person?

For knowledge utilities to improve our commonwealth of nature, then first, practical, technical, and social requirements must exist in the physical world, which then require the collaboration of both public and private organizations to build and maintain trusted software and hardware infrastructure. The distributed knowledge infrastructure does not exist without real resources and costs. Servers, fiber optics, and satellites all require power, maintenance, and agreements on how they all will work together. That requires layers of agreements on common language and structures that are codified in software. There is a loose collection of organizations that already steward some of this core knowledge utility infrastructure. These stewardship roles are not served by nation-states but by a network of community governance organizations that manage the functioning of the internet. For example, the Linux Foundation, Apache Software Foundation, and Mozilla Foundation currently serve as stewards of key software infrastructure. To meet the needs of a robust and trusted commonwealth of agricultural knowledge for creating shared abundance, these types of enabling non-nation-state governing organizations will need to be expanded and strengthened.

Abundant Knowledge

What is scarce? Surely time is scarce? This is true in the sense that we get only one life, but yet again there are ways in which competition and how we use our time can make us feel an artificial sense of time scarcity. Each time we are able to build on the work of others with confidence, each time we use the elements of life pulled from our commonwealth of agricultural knowledge, we bundle time, and so get the benefit of multiple lifetimes. Each time nature uses genetic code that has been developed over millions of years, millions of years of development are collapsed into something that works in our lifetimes. Each time we add to that collection, we are putting our lifetimes' work into a useful form for the benefit of future generations. At the same time, yes, we each have only our own single lives in which to pursue happiness. The goal is to spend as much of that time in a framework of sharing abundance rather than having it squeezed into a life of scarcity and competition.

In contrast, we need not look far to find lots of frustrating examples in which our time is treated as abundant when we would rather have it be valued as scarce. It happens each time we must stand in line at the DMV, fill out redundant forms at the hospital, reproduce others' efforts by spending time searching for knowledge or data that already exists somewhere, create a report that no one reads. In those cases, we are creating and living in artificial and unnecessary time scarcity.

Time is indeed one of the most curious elements of life, especially since our lifetimes and those of plants and animals all move at different rates. We know, for example, that the urgency to address climate change is really on our human scale, not geologic scale. The Earth has been through greater upheavals and mass extinctions and will likely go through them again, but for the narrowest of narrow bands of human history on Earth, we require very specific conditions for us to continue to thrive as a species. To keep our planet within a habitable and abundant balance, we have, as Howard Buffett noted, only "forty seasons" to learn and adjust. That is why building on one another's work is so important. One farmer can have the benefit of forty seasons and pass some of that experience down, but if 1,000 farmers do the same, there is the collective benefit of 1,000 years in a single year. If a million people participate, then a million years of collective experience are available. If we are then able to compound knowledge across generations and deepen our understanding of human and natural history, we add even greater richness. It is in this way of bundling our experiences for continual improvement, with compound interest, that time shifts from a scarce resource to being far less of a constraint, if not truly abundant. However, for time to be compounded, knowledge must be shared, and real resources, energy, and infrastructure must exist and function to support and grow our commonwealth of knowledge.

What are some principles for creating abundant knowledge? Again, we come back to governance of the commons as providing key guidance and many design principles that build nuance and detail around the implementation of agroecology. There are many branded philosophies about how to manage communities; they have names such as Pro-Social, Sociocracy, and Collective Impact. These approaches focus on collaborative governance and the social dynamics necessary to support collaboration across organizations as well as cultural boundaries across different scales. Other approaches, such as holistic management and permaculture, focus more on the relationships between humans and nature at the farm or small-community scale. These approaches converge in agroecological principles that bridge the management of both larger-scale human communities and local biological communities. To be made actionable, principle must be translated into plans and practices. These must also feed back into a cycle of adaptive management and continual improvement so that each version can build on the previous. As we look for abundance, a number of assumptions about how people work together also change.

We collectively have what some have called a global cognitive surplus, or the collective capacity for innovation that we can share without diminishing our supply. In terms of managing and visualizing such a complex system, open-source software is a great repository of the cumulative, multigenerational knowledge and unifies and stores our understanding very much like the universal language of logic like mathematics. While theorists can write copious papers about how we think the world works, software is our best tool to actually describe the complex interactions, observations, and descriptions of "what works where." An agricultural knowledge commons is built upon our understanding of natural systems, which is then embedded in code, which is then embedded in the latest version of the software we use, all of which is improved and refined over time.

However, software also has the power to edit our experience and create "walled gardens" and blinders to limit knowledge through paywalls or through algorithms intended to limit the field of view and spoon feed content intended to influence and control people, rather than to provide context and illumination. We are still collectively working to shake off the blinders of the extractive industrial model of development and knowledge sharing as discussed in chapter 3. Fortunately, the beauty of software is that it can be copied in full fidelity infinitely and without degradation while also evolving in real time. Like any language created by a community, software is a reflection of the people who created it. Software can be rigid and unforgiving when created with power and control in mind, or empathetic and welcoming when the intention is reciprocity. As a language, it is unique in human history as a method to rapidly compound our shared knowledge and begin to comprehend and manage complexity.

Public Land Library

An initiative lead by OpenTEAM imagines a new kind of knowledge utility called Public Land Library, which is a place where participants can both contribute and curate specific knowledge that maintains its integrity, attribution, and accessibility. The library comprises a curated set of existing and new environmental, geological, and human records of land that is searchable like a registry of deeds. The technical term is a cadaster, which represents a comprehensive recording of the real estate or real property's metes and bounds, often represented as a map.

The ambitious concept of Public Land Library has deep roots in previous efforts to not only map the soils of the world but also the seeds, biodiversity, human history, and local knowledge so that they can be shared. So many previous attempts have been made to achieve this vision. There is a direct lineage of efforts:

- Diderot in the 1700s
- USDA extension and land grant university system in the 1800s
- agrarian movements in the early 1900s
- FAO's early efforts to support reconstruction and development through land and soils mapping, agricultural-knowledge clearing houses to facilitate local market development, and the redistribution of land in postcolonial independent nations following World War II

These early efforts were motivated by a desire to reduce global inequities and increase food security and sovereignty. However, they were hampered by centralized approaches and huge bureaucratic burdens as well as the challenge of organizing mountains of data, without computers or the internet, in such a way that would be detailed enough to be useful at a local level.

Based on this cycle of development, the timing is ripe for another attempt at this knowledge utility envisioned by so many before, but this time with the benefit of new technology and distributed participatory approaches that may avoid many of the previous shortfalls. OpenTEAM and other projects like it represent a new attempt to realize these ambitions, which also encompass many other organizations, across the United States and internationally, such as the OpenGeoHub Foundation, Open Earth Foundation, Buckminster Fuller's World Game, Digital Green, LandPKS, Practical Farmers of Iowa, California Farm Demonstration Network, Quivira Coalition, and Grassfed Exchange. Individually and together, these organizations are creating new and innovative localized structures for shared learning.

Previous attempts to create Public Land Library faced several barriers to their effectiveness, key among them that data is not the same thing as knowledge and that knowledge must be applied in local contexts to be useful. Until recently, mapping at a high enough level of detail to be useful has been extraordinarily expensive when done through a centralized approach. There is tremendous complexity in navigating the biological, technical, and social contexts for any land steward to access and use these tools. Libraries generally do not function well without librarians, and so our Public Land Library also needs skilled facilitators and trusted digital intermediaries to support the public nature of agriculture as a shared endeavor, which does not entirely rest on the shoulders of individual land stewards.

The biological equivalent of a digital intermediary is the relationship of fungal networks with the roots of plants. Although easily missed, these microbes provide a critical ecosystem function to translate between the systems and to distribute resources. In forest and agricultural systems, a lack of the intermediary "mycelial" function dramatically effects the resilience and growth of plants and humans alike. In the United States, the digital role has primarily been called *technical assistance* (TA) and involves trained facilitators called extension agents. The second director general of the FAO in the early 1950s, Norris E. Dodd, said, "By an extension worker, I mean a man who lives among the farmers in a county or district who has their confidence because he helps them, who knows enough about the practical agricultural problems . . . to really help the farmers, and who isn't afraid to get his shoes muddy and his hands dirty with the job."[1] In India, the role is often called a *village resource person*, and in Australia the role is organized within Aussie Farmers Mutual, an organizational structure based on historic agricultural cooperatives.

The FAO in its first decades imaged a global agricultural extension service equipped with appropriate site-specific knowledge to reconstruct and rebuild newly independent nations around the world. It is an idea that formed in the wake of the Dust Bowl period of the 1930s, and it has echoes of land grant colleges, agricultural experiment stations, conservation districts, and cooperative extension service providers. But while these efforts had ambitions for a knowledge circulatory system, in practice the attempts were more like pipelines running one way. In contrast, Public Land Library is a non-nation-state–based structure to enable a reenvisioning of land-based knowledge for the twenty-first century. It's a new approach to using decentralized collaborative technology and governance and will enable the radical updating of our agencies, institutions, and intermediaries so that everyone may be equipped with the useful knowledge that's necessary for implementing the Great Regeneration.

Data and Food Sovereignty

The tension of having publicly accessible knowledge is in simultaneous importance with maintaining local control and identity within a global system. Previous efforts, due to technology and governance constraints, often resulted in centralized control, and so, reduced trust and local autonomy. Those tradeoffs of global knowledge and local control are no longer necessary.

Tim Berners-Lee, an early contributor to the development of the World Wide Web, developed an important critique of the way internet communication currently works, and of the unfortunate domination of tech giants fueled by vast troves of centralized data, which is, in effect, *our* data. We are well aware that these tech giants have become surveillance platforms; Berners-Lee, and many advocates for what is being called Web 3.0, believe the tech giants have become the primary gatekeepers of innovation, as well. He offers an alternative approach to technology that gives individuals more power.

The idea is that each person can control their own data—such as records of websites visited, credit card purchases, workout routines, music streamed—in an individual data pod. Sharing a person's data with a friend, a doctor, a researcher, or a company with permission can be managed through a secure link for a specific task that enables users to link to and use personal information selectively but not to store it. This is a radical alternative to a world where internet users are accustomed to accepting cookies and allowing web crawlers access to their personal information through coercive and unreadable data-use agreements. The PodBrowser concept, as his company calls it, flips how we think about data on its head. In this model, we provide data only for the purposes we intend. This change puts the individual in control and creates transparency in how data is used and thus forms a renewed foundation for trust. This has huge implications for science, research, and even tracing the provenance of digital artifacts such as photos used in the media. We are seeing the demand for some of this same kind of function in the proliferation of non-fungible tokens (NFTs) that use distributed ledger technology.

OpenTEAM and the Ag Data Wallet

Is the unique, place-based sovereignty and independence of local agriculture compatible with a global agricultural knowledge commons? I believe the answer is yes. Indeed, local control, the improvement of individual freedom of expression, and the local enjoyment of abundance are the ultimate ends of knowledge exchange. But how does this work in practice? How do we balance the tensions between global knowledge and local sovereignty and control?

Several new kinds of knowledge utilities are now being developed to create trails of attribution and trust necessary to balance these concerns. These efforts have culminated in the creation of an Ag Data Wallet by the Open-TEAM community, with contributions from other nonprofit technology organizations such as Digital Green, AgStack Foundation, Linux Foundation, Tech Matters, Our Sci, farmOS, Regen Network, Hylo, and Terran Collective. Just as a real-life wallet holds multiple cards used for different purposes, the Ag Data Wallet holds a unique identity and agricultural data set, each with unique permissions that can be easily transferred to different end users to achieve different goals, while the wallet holder remains in the control of data.

The word *wallet* evokes both a place where important documents are kept and something that is under a person's control. By creating and securing individual identities and giving users the ability to delegate authority to intermediaries, an individual's data could be entered once and used many

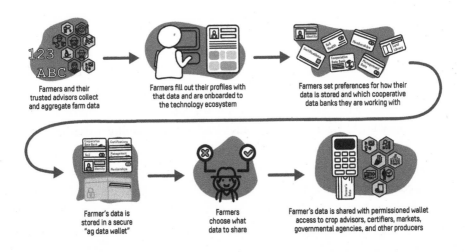

Farmers and their trusted advisors collect and aggregate farm data

Farmers fill out their profiles with that data and are onboarded to the technology ecosystem

Farmers set preferences for how their data is stored and which cooperative data banks they are working with

Farmer's data is stored in a secure "ag data wallet"

Farmers choose what data to share

Farmer's data is shared with permissioned wallet access to crop advisors, certifiers, markets, governmental agencies, and other producers

times while maintaining individual privacy, autonomy, and sovereignty. The ambition to enable people to "enter data once, use many times" might sound quite simple, and yet, that concept results in a radical realignment of the way we envision how the digital world should work. It is an important concept and vision because achieving it also requires a change in how we relate to one another and how and for whom we build technology.

Land stewards and groups are able to use an Ag Data Wallet to assure trust. They also enable peer-to-peer benchmarking, the organization of community discussions, the creation of new markets and financial tools, and access to planning tools all through a growing suite of interoperable open-source software. Socially, this technology also enables the creation and enforcement of what might form a prototypical Ag Data Bill of Rights:

- Data can be used only for the specific task the user authorizes
- User controls rights for:
  - which entities (if any) can access data
  - what uses that data may be shared
  - how (and whether) that data may be aggregated
- User has the ability to:
  - revoke rights for any individual entity or specific use
  - download data from the wallet platform in a common, portable format
  - remove data from the wallet platform
  - assign proxies to act as fiduciaries on their behalf

A key part of this system is interoperability, or the ability for multiple tools to connect and exchange information. Allowing users to enter data once and use it for many different purposes is fundamental to creating a technology ecosystem where different tools work together to enable people to easily collaborate and share data across networks and organizational boundaries. The function of an Ag Data Wallet expands a land steward's ability to manage their own data and not cede control to monolithic tech giants for convenience's sake. Unique and trusted identities also provide automatic attribution for data contributed to the agricultural knowledge commons. As new tools like the Ag Data Wallet, Public Land Library, and other crucial knowledge utilities emerge, technical service providers and digital intermediaries will play an increasingly important role. Digital intermediaries are equipped with the tools that support both trusted relationships and access to knowledge.

The Digital Agrarian's Coffee Shop

Imagine if you could walk onto a field with limited training and in five minutes have an assessment of the current health of the soil, roots, and biota of your land. You could benchmark not only available options to improve productivity, diversity, and profitability but also then connect with others with similar situations to discuss those available options for improvement. That science-fiction story became science fact in 2022, and it began to fulfill visions of knowledge exchange and technical assistance dreamed of by early leaders of the FAO and Indian economists such as Samar Ranjan Sen. However, there's one key difference. This new approach is not a centralized repository of expert knowledge; rather it is participatory, decentralized, and facilitated with expert technical assistance and open-source software.

Planning conversations over morning tea or coffee is a common phenomenon across the world. Coffee shops and salons have been hubs for the exchange of ideas and information, as well as political innovation, for centuries. In farm country around the world, local coffee shops are where farmers and ranchers share notes about who is doing what, report on how fields are looking, and tell stories about weather and equipment, because data and observations become more useful knowledge with context. Powerful open-source analytic and communications tools are now all being wrapped into applications that are attempting to replicate a part of the social experience of knowledge sharing that happens in those important local institutions. The difference is that these technologies enable people to gather and find one another regardless of where they are located on Earth. The opportunity for sharing context across geographic and cultural boundaries and benchmarking local observations to others regionally or globally is represented by a project called the Farmers and Ranchers Digital Coffee Shop benchmarking tool also being stewarded by the OpenTEAM community. The creation of a digital commons and coffee shop is a first version of our agricultural knowledge commons made visible at an individual level. The Ag Data Wallet forms the basis, to extend the coffee-shop metaphor, for land stewards to choose a private booth if they would like to talk about more sensitive topics with only a trusted few. Or, if they want, they can sit at the counter and brag to the whole world.

The Digital Coffee Shop was created so that the complex variables of soil, management, and land are visualized in the local context and build on not just the other knowledge utilities but also the powerful shared logic of mathematics in open-source statistical tools and powerful graphic visualizations that, in turn, help create new data-driven stories. This effort is being led by Our Sci, a Michigan-based, open-source business that has a mission to support community-driven research. Our Sci grew out of large-scale, distributed public science efforts such as the Real Food Campaign, Gathering for Open Ag Technology (GOAT), and the international Gathering for Open Science Hardware (GOSH), which itself emerged initially from one of the largest open science projects on Earth at CERN in Switzerland.

Tools like the Digital Coffee Shop benchmarking tool enable individual observations, which on their own are often uninterpretable, to be placed into the context of others', and they enable farmers and their advisors to form groups where they can compare and contrast their results. They are then benchmarked with others of the user's choosing. It is collaborative tools such as these that help filter out from our vast shared global commonwealth of agricultural knowledge what is actually useful. It enables individuals to walk through options not only with their neighbor but anyone with whom they choose to engage. Those conversations are made possible because of open-source social media platforms such as Hylo, a product of the Terran Collective created as a free and open-source alternative to Facebook. Hylo, on its own, is a communications tool centered on place, self-organizing groups, and shared resource discovery. It is an example of how we can ask questions, work together, and share with the potential to create value.

Knowledge utilities like the Digital Coffee Shop benchmarking tool, Public Land Library, and Ag Data Wallet are directly tied to freedom in the sense that they open up options and possibilities. Free technologies and services are not all free as in "free beer," as the Free Software Foundation likes to point out. All that free sharing of knowledge is transacted with a cost not just in terms of labor but also in physical materials and energy. What are some of the core elements necessary to provide knowledge abundance and to build the fiber optics, processors, storage, and other shared functional essentials of knowledge utilities? In the previous chapter we looked at the elements of biological abundance at a global scale. If we apply the same quantitative analysis to the elements necessary to build and power the knowledge utility infrastructure, we find a similar pattern of potential abundance but with unique qualifications.

Silicon, Lithium, and Aluminum

A portion of a knowledge utility exists as ideas captured in open-source software—an abundant, nonrival resource. But, as mentioned, that open sharing has a cost. The information we transmit in a knowledge utility is nonrival, but the energy and raw materials that go into constructing and maintaining the infrastructure require that we make a choice not to use those same resources in other ways. What materials will those portions of the utilities be composed of, and what kinds of substitutions are possible? What do ubiquitous digital communication and open knowledge pathways run on?

It is beyond the scope of this book to do an exhaustive analysis of all the elements that are required to build the infrastructure to support knowledge utilities. However, a few essential components, such as silicon, aluminum, and lithium, stand out. Silicon is used for fiber-optic glass, processing, imaging, and storage chips; solar panels are built on aluminum structures; and we use lithium-based power storage, with even more abundant substitutes on the horizon.

Innovation and information (free and nonrival) come with the ability to re-create ideas and colors with perfect fidelity. Such replication is made possible because of silicon-based transmission, storage, and processing. The availability of silicon for devices and fiber optics used to transport, store, and power ideas for every person on the planet does not need to be a limiting factor in the development of the infrastructure to support a ubiquitous knowledge utility; raw materials are abundant, even as chip plants are currently concentrated in just a few areas on Earth. In 2022, the vulnerability of supply chains to pandemics and conflicts, such as the Russian invasion of Ukraine, brought world attention to the concentration of chip manufacturing and other key processes in countries such as Taiwan. By incorporating cradle-to-cradle product design, the potential shortfall of rare elements in integrated circuits can be largely addressed. With abundance as the mindset, products can be better designed to be upgraded and upcycled rather than to be discarded into landfills.

Aluminum and lithium are of particular interest, because of their relative abundance and availability, to build and govern the knowledge utilities. Although lithium is widely distributed on Earth, it does not naturally occur in elemental form due to its high reactivity. Lithium constitutes about 0.002 percent of Earth's crust, making it less abundant than the other minerals we have discussed, but still not rare. In particular, lithium forms a minor part of igneous rocks, with the largest concentrations in granites. Granitic pegmatites also provide the greatest abundance of lithium-containing minerals. Worldwide reserves identified in 2021 were estimated by the US Geological Survey at 21 million tons, with 75 percent of all global lithium occurring in the ten largest land deposits. However, the accurate estimate of world lithium reserves is difficult due to the different forms in which it is found.

Silicon is the eighth most common element in the universe by mass. More than 90 percent of the Earth's crust is composed of silicate minerals, making silicon the second most abundant element in the Earth's crust (about 28 percent by mass), after oxygen. If the Stone, Bronze, and Iron Ages were defined by the metals used in everyday life, we might say we live in the Silicon Age today. When we walk on a sandy beach looking out at the ocean, we might imagine the sand beneath our feet (composed of silicas primarily) is the place where the mineral begins to be spun into a global digital nervous system connecting people to people and people to nature in new ways. Amazingly, silicon also makes it possible to capture power from the sun in the form of photovoltaics. Pure silicates are expensive to produce, not because of the materials themselves but because of the processing required to make them useful. There is an elegance to the feedback loop of silicon being used to create and store thermal power that can be further used for processing or recycling or moving water or many other endeavors, but also to repair and support tools, enable communication and innovation, and create a parallel nervous system to the carbon one already in place. If that wasn't enough, silicon compounds are incredibly heat resistant, with boiling points of $1,414°C$ to $3,265°C$, the second-highest among all the metalloids and nonmetals, making our silicon creations durable, stable, and dependable.

The lithium-ion battery market is growing rapidly to meet the needs of transportation, the construction of power grids, and a rapidly expanding array of innovative tools and devices. I have observed this trend firsthand on my farm, and now prune field edges and clear fence lines with lithium-powered chainsaws and equipment, where I might have used a gas-powered saw just a few years ago. The quiet, odorless power is remarkably welcome and makes the jobs even more enjoyable. With the increased demand for lithium-ion batteries, understanding the relative abundance and recycling potential of lithium is important to our view toward scaling regeneration infrastructure, recognizing that the next battery technology may be based on even less-rare elements.

The total lithium content of seawater is estimated at 230 billion tons, where the element exists at a relatively constant concentration, though much higher concentrations are found near deep ocean hydrothermal vents. Previous estimates put the land-based supply of lithium, not taking into account the discovery of new sources or recycling at the current projected use, at 300 years. But at the pace of current adoption of electric cars and batteries, the known supply will be drawn down far more quickly. Projections for large-scale electric power generation and transportation systems also often assume a recycling rate between 80 to 90 percent based on an industry that is yet to fully develop and which is in the low single digits today. Recent EU legislation is clear: For every lithium-ion battery brought into circulation, it is the manufacturers' obligation to deal with its afterlife.

That's where the term *recycling efficiency* comes in, meaning how much material can be recovered. A 2020 paper from the Royal Society of Chemistry (RSC) described the difficult task of collecting and recycling lithium-ion batteries.[2] The lack of any standardization of cells and the predominance of cells from small portable devices mean that initial recycling approaches will be similar to solid municipal waste, producing streams of lower purity. The homogenization of cell design and chemistry, and the larger fraction of similar automotive cells, will enable easier recycling with streams of higher purity and higher value. From a "green chemistry" perspective, the RSC authors remark how it is the scale of market growth that necessitates the manufacturing and recycling process to be as efficient as possible. This seems reasonable, given that lithium can be recycled and is of crucial importance to the future. As such, we can put lithium in the abundant category as a transition mineral. Furthermore, it is certainly possible that a clever process for concentrating lithium from seawater will be developed (perhaps by using abundant silicon-based solar energy and biological pathways from our biodiversity tool kit). If lithium can be recycled and recovered from seawater, it can safely be moved into a "forever" supply in the scale of human history.

If lithium-based power storage becomes a bridge to hydrogen-based power storage, the scarcity illusion vanishes even further.

Aluminum is widely used in the digital economy, including in HD TVs, smartphones, laptop computers, monitors, digital cameras, and game consoles. It is lightweight, durable, noncorrosive, versatile, and conducts electricity easily. It is found abundantly in the Earth's crust, where it binds to rocks as a result of a chemical reaction with oxygen. Smelting the raw ore and separating out the aluminum is done by an electrolysis process that requires a great deal of electricity. If not sourced from renewables, the carbon footprint of producing aluminum can be very high, which is why many smelters are located near abundant hydro or geothermal power sources. These high costs have encouraged a considerable recycling industry, though the effort to reconstitute commercial-grade aluminum from scraps also requires electricity. I include aluminum as an example of how a regenerative ecosystem thrives on systems that are good at creating communities and recycling nutrients within them. While the process of recycling aluminum may be more demanding than most will want to attempt at home, nearly anyone can create a backyard foundry to melt down aluminum scrap and cast it (there are plenty of how-to videos on this). A neighbor of mine melted down an old Nissan aluminum-alloy manifold to sand-cast (another use of silicon!) cylinders for a motorcycle he fabricated himself. So even the recycling technology is not inaccessible at the small scale, though the recycling efficiency of a backyard smelter is certainly low.

Now what about rare earth elements and chip shortages currently limiting supply? At present, the predominant use of rare earth elements such as neodymium and dysprosium is in the permanent magnets of offshore wind turbines. Some onshore wind turbines also use them, as is the case for turbines in about 3 percent of wind farms in France, but alternatives exist. For example, it may be possible to make asynchronous or synchronous generators without permanent magnets to reduce the need for rare earth elements. But without alternative solutions over the next ten years, the wind sector may end up accounting for 6 percent of annual neodymium production and more than 30 percent of annual dysprosium output. There are clearly issues with distribution, recycling, and more, but the bigger point is that these elements are not actually all that rare, they can be recycled, and often substitutes are possible.

Physical infrastructure is like a skeleton and nervous system without a body. To animate the body, the operating system—the software that makes the body purposeful in motion—must also exist, as the brain. And that perhaps is where we have seen some of the greatest progress in the last decade. The real expression of this evolution of course is not the hardware or software but the form that emerges when they are meaningfully combined. In the process of the Great Regeneration, the combined form of hardware and software gives a voice to the commonwealth of nature.

Silicon Gives Voice to Nature

François Quesnay, an eighteenth-century economist and physician, was the first person to associate the circulation of blood in the body with the flows of products and money among farmers, landowners, merchants, and craftsmen across the landscape. His thesis, first described in the *Encyclopédie* of Diderot, argued that there are *real flows* (the circulation of goods and flows of nature) and *financial flows* (the circulation of money). The two flows relate to one another but, until now, we have not been able to map financial flows very well to the flows of nature.

We share this planet with an estimated 1.5 million nonhuman animal species. Only two million of them are recorded by science. Yet we scarcely account or give voice to how they move through the world. There has not been a way for animals, plants, other kingdoms, or the ecosystems upon which we all rely to make themselves known to us online or to express their preferences to us or to hold assets and currency or to have legal standing and representation. However, our new digital observatories can give voice not only to plants and animals but to whole ecosystems, including salt marshes, estuaries, rivers, spawning grounds, and migratory pathways. How might we give voice and agency to these entities in the landscape and at a global scale? How might we give economic power in the system to these important nonhuman actors upon which we depend to co-create regeneration and to assure that they are both healthy in their own right and, selfishly, healthy for the sake of our own prosperity, well-being, and enjoyment?

As we make decisions on our own behalf and find the boundaries of what we manage, while also seeing the bigger picture, we can also see the boundaries and dependencies that are not represented. For example, across OpenTEAM hubs of farms and ranches on the Chesapeake Bay, there are countless unvoiced participants and nonhuman actors—the estuary itself, its tributaries and the fisheries, and the human and nonhuman communities that are dependent upon it. These patterns of nature by nonhuman actors are part of the ebb and flow of a living landscape and play into the success or failure in the biology, as well as the economics, of an individual farm working to improve the conditions for the Chesapeake estuary, rivers, and streams. As an actor in the economy, a watershed certainly represents more influence in our economy than any individual human, or even a large corporation. And yet without representation, a watershed cannot engage in governance or influence policy and must instead rely on our haphazard advocacy on its behalf.

We have created corporate entities and trusts that have legal rights, just as people do. They can hold assets and have legal standing to facilitate continuity and function that transcends individuals. Often, we think of the corporate structure as being extractive of nature and people, and yet the structure itself could play the opposite role, representing people and nonhuman entities' role in the community. If we didn't have the concept of a corporation, how might we invent a structure that would then represent the continuity of entities that clearly have an important role in our resilience as a species and in our well-being? For the commonwealth of nature to exist and for farmers and ranchers to profit from its improvement and transactions, it must be made visible. The agricultural knowledge commons forms the foundation for a silicon nervous system. This nervous system in turn gives presence and sentience to nature, which can then be made visible with data and data-driven stories we can share.

One of the unexpected attributes of Ag Data Wallet and Public Land Library is that the architecture is not limited to only human identities participating in transactions. Nature can represent itself and provide and award a claim of positive or negative contribution directly to other entities. This is no longer just theory. In the fall of 2021, the New York Stock Exchange and Intrinsic Exchange Group launched a new asset class called Natural Asset Companies (NACs), creating a new market whose assets generate trillions of dollars in ecosystem services annually.

Corporate personhood creates a meta-organism in our legal system that has some of the rights and privileges of an individual; however, corporations are not mortal and can outlive individuals as they carry out their defined mission over time. In this sense, nature-based corporations, if they were to exist in large numbers, could become persistent fiduciaries for nature's intrinsic value. Countries such as New Zealand have led with the creation of rights-of-nature corporate personhood. The Regen Foundation, which was formed in 2020, is granting currency and governance rights to rights-of-nature organizations, which represent rivers, salt marshes, mangrove forests, mountains, grasslands, and lakes from Australia to the Amazon.

These organizations could in turn contract to directly verify nature-based claims of other corporations even over many human lifetimes. Land and the attributes of land, including biodiversity and changes in soil over time, could be represented in the agreements and contracts for ecosystem services that are already being drawn up by markets. Markets require two entities, not necessarily human entities, to enable a transaction on agreed-upon terms. When it comes to environmental services, the terms of the agreement must be based upon the ability to make a claim about environmental change over time. A Public Land Library and an Ag Data Wallet provide the kind of high-quality data that the commonwealth of nature requires to be represented in human-readable ecosystem service contracts. The land library is a public record of what exists and what changes over time, and the wallet enables individual land stewards to attest to what they did as individuals to contribute to that change on the land.

Within the last few years, even corporate giants like Walmart, PepsiCo, General Mills, Danone, Mars, and Unilever have made environmental commitments related to both climate and biodiversity. It turns out that setting science-based environmental targets is also good business. Stocks of corporations with science-based goals vastly outperform those without, even without transforming current short-term market incentives. Since 2006, when the certified B (Benefit) Corporation designation was launched, more than 4,673 B Corps across 155 industries in 78 countries have been registered. Although not at all an extraordinary concept looking at a longer human history on the land, it is remarkable that it has taken Western industrialized cultures this long to recognize our commonwealth of nature as a full participant in our legal and economic system. Agriculture could be seen as a fiduciary role to the land we steward and temporarily hold in trust.

A proliferation of environmental markets and certifications is following a similar trajectory to create value from our shared public land libraries and related knowledge utilities. One example is the Ecosystem Services Market Consortium (ESMC), which was founded in 2019 and includes some of the largest food companies in the world. Other private markets include efforts by Hudson Carbon, CIBO, NORI, Regen Network, Nutrien, Indigo Ag, Cargill, and many more. The USDA is also taking significant steps to create a system of verifiable and credible environmental improvement claims related to what they call Climate-Smart Commodities. They allocated $1 billion to begin seeding the effort in 2022. Ultimately, these systems must deliver on improved soil organic matter or other environmental services. There is no gaming the underlying natural system—ultimately, nature cares not at all for our legal constructs or claims or market theories. However, these human contrivances of theories, stories, markets, and contracts are created for our benefit, not for the soils'. The real proof will be if these innovations illuminate our commonwealth of nature and subsequently align human behavior with improved outcomes over time.

The OpenTEAM-led Public Land Library effort is being leveraged by land stewards, amateur scientists, governments, and non-governmental agencies (NGOs) to help design a governance system that will be trusted. Others can contribute to create a shared public record of past activity and the current baseline of conditions for land anywhere—from an urban backyard plot to a million-acre ranch. This is a radically different approach from previous ambitious, but unsuccessful, global mapping efforts, which attempted to create a professionalized, centrally managed workforce to take on this task over decades without the assistance of a global soils knowledge utility. A distributed, interoperable open-source approach enables the public tracking of change over time not just for public lands but also the voluntary mapping of the natural state of all urban and rural lands. By placing observations into the public domain, land can be searchable, like a digital county registry of deeds. Sensitive data about land, like protected species, can also be managed using the same tools. Sovereignty and control are the key words here. This capability to control access to knowledge is now made more viable by using distributed ledger technology. Increasingly this technology is not limited to claims for markets but also claims for origin, nutrient density, or social or environmental data related to the nonfungible value of products to a purchaser.

OpenTEAM members are curating key pieces of this shared infrastructure, including the Purdue Open Ag Technology Systems Center's (OATS) Trellis project, Regen Network's ecological ledger, Digital Green's FarmStack, and the Linux Foundation's AgStack project. Each of these efforts represents a key piece of a new infrastructure that is governed not by nation-states, but is being collaboratively built, out of necessity, to support a new approach to knowledge systems, to enable new kinds of markets and clearinghouses, and to build trust without limiting local adaptation and innovation.

CHAPTER 8

Soil, Silicon, and the Great Regeneration

The ultimate, hidden truth of the world is that it is something we make,
and could just as easily make differently.
—DAVID GRAEBER, *The Utopia of Rules* (2015)

Imagine if you could instantly connect every farm to the history of its land and region, allowing you to see not just the direct value of the food, fiber, and fuel the farms produce but also their commitment to environmental stewardship and their social role in the community. Imagine further that on top of being able to explore the origin of the food, you could instantly assess the nutrient density differences in your food purchases. Also imagine if the farmers and ranchers who helped make the product could just as easily share practical knowledge with customers and other producers in Argentina, northern India, Australia, the United States, or anywhere on Earth. One more step—imagine that these farmers and ranchers and their communities could be recognized and rewarded for their contributions to accelerating regenerative outcomes. Now, imagine if parts of this vision are already happening today—that the Great Regeneration is already underway.

In this Great Regeneration, land stewards are connected to the world beyond their fields and forests. The concept of the lone hero-farmer has always been a myth. Agriculture is a shared endeavor, with the transformation of soil being a product of generations of observations, testing, and sharing on what works where. Let's explore how this works by looking at a relatively new farm located on the Chesapeake Bay.

Amanda Cather moved to Maryland and established Plow and Stars Farm in 2014. She grew her operation from a few sheep to a 200-acre diversified farm in just a few years. The grazing pastures include a diverse mix of blue-grass, white clover, plantain, and buckwheat. She has noticed that as the soil in the pastures develops, the plant and insect ecosystems become healthier, and the sheep eating those plants grow stronger and have more robust immune systems. Quantifying and sharing observations of these kinds of changes, which happened in a relatively short time, has fascinated her since she started farming.

In 2017, Cather began using the Cornell Assessment of Soil Health, which guided her early efforts and inspired her collaboration with others who were also using the same assessment system. Her approach to sheep farming and cropping systems is an example of the biological, technical, and social feedback gained by benchmarking her soils to those other farms and then sharing that data with the Cornell Soil Health database. That simple act also illustrates the potential for accelerating the development of local knowledge and global learning. Cather regularly consults with her neighbors (many of whom also exchange knowledge extensively with the local soil and water conservation district) in choosing pasture mixes and exploring cover crop varieties. She also exchanges knowledge within several more formal net-works, such as the Million Acre Challenge, which she cofounded and which is also a member of OpenTEAM as part of larger collaborative agricultural hubs and networks that extend nationally.

Through this collaborative network, Cather is able to connect with the more than forty organizations representing farmers, marketplaces, research universities, food companies, agricultural technology companies, and non-governmental organizations that all help share knowledge to improve soil health. These collaboratives structure the accumulated observations of individual farms like Amanda Cather's. This accumulation then helps her create plans that are based on soil health tests and shared field observations from across the country. How should she decide what might work best for her operation? Cather, like each of us, has perhaps forty to fifty seasons to learn and experiment during in a lifetime. What may seem like such a simple question—"What should I plant this year and where?"—can result in a complex set of interactions ranging from consulting a neighbor to tapping into a global knowledge utility for data-driven guidance. Although Cather's farm is but a small piece of a much bigger agricultural mosaic, her work is simultaneously amplified and amplifying within the vast network of inter-connected agricultural practices and knowledge as she contributes to, and receives from, the agricultural knowledge commons.

We can see this play out in the work that Cather has done as a founding member of the Million Acre Challenge, which has formalized the process of regional knowledge exchange. A group of Maryland farmers formed the Million Acre Challenge collaborative 2020 with the intent to build soil health, increase farm profitability, and improve water quality while making farms resilient in the face of climate change. The farmer-focused collaborative uses soil health science, economics, education, and incen-tives to achieve the vision of enhanced soil and ecosystem health, as well as increased farm profitability, on at least one million agricultural acres in less than ten years. The goal is to build bridges between an individual's farm, such as Cather's, with others across the Chesapeake Bay watershed. While Cather and each of the farms in her network of the Million Acre Challenge are making independent decisions, they are also drawing on col-lective knowledge and collaborative tools. Through this work, participating farmers such as Cather are building a framework that contains a suite of options from questions like:

- What soils will she be planting into?
- What varieties of seed are available and what works well with one another?
- What crops preceded it and will follow?
- What is the climatic zone and the projected weather pattern?
- What is the effect on biodiversity and the ecosystem in which she lives?
- What are the short- and long-term trade-offs and returns?
- What method of planting and harvest will be used?
- Will grazing animals be incorporated into the system?

These questions shape which decisions are made. All have nuanced answers based on the local conditions, which can help determine what combinations will work where. This process is the product of personal relationships and cultural exchange built over generations by people living and observing the patterns of nature closely. However, we now have an urgent need to bring on new land stewards to help manage these small details of nature in the shadow of climate change and global demands on resources and biodiversity.

Fortunately, new decision-making tools are becoming available to accelerate learning. Even under normal circumstances, a skilled agronomist may struggle to think through the variables in a coming growing season and to select the best options for cover crop varieties. Under the unprecedented conditions brought on by climate change, this job will become even more challenging. But with digital tools and access to years of observations, Cather and her neighbors can quickly walk through the available choices. That is precisely what is now available from the Northeast Cover Crop Council's Species Selector Tool, driven by the plant data service that a special kind of agricultural knowledge utility prototype created out of Purdue University's Agricultural Informatics Lab and Open Ag Technology Systems Center (OATS). Dr. Ankita Raturi and her lab are building not just a plant data service but also related tools in collaboration with the AgStack Foundation and the Linux Foundation to make weather, climate, soils, and input knowledge universally available. Parallel governmental efforts are simultaneously being launched. A USDA program, for example, aims to create data services that will enable producers to access maps of dynamic changes in soil health, and to make decision and planning tools free and open source. Access to an agricultural knowledge commons in all its forms is necessary but not sufficient for success in the field.

Emerging regenerative patterns and practices are not all found or used on any one farm or ranch. Every enterprise is different, and every landscape is unique, but taken together they form a pattern. There is no one type of regenerative artist. The detailed patterns of sunflower, wheat, corn, chestnut, peaches, rhubarb, and grazing on my farm are already different from those of my childhood, and will be different in fifty years from now, just as two chefs will create a different meal from the same ingredients. We each start with a common palette of biodiversity from which we work individually with the elements of life to create green, black, brown, and blue patterns through our stewardship. A rancher such as Emily Cornell of the Sol Ranch in New Mexico is a Rembrandt of land stewardship. She mixes her palette on the land each year with her herd management, forest thinning, riparian restoration, and prescribed burns. After years of painting detailed patterns in concert with nature, a complex mosaic of intricate patterns emerges as a dynamic masterpiece that can be experienced and admired both on the ground and from above. Cornell stewards a process that creates a living, breathing, and edible landscape of colors and flavors for animals, birds, insects, and humans to enjoy. Unity of purpose emerges when we step back to observe the larger patterns we are creating together.

As we place the boundaries of our land in a broader context, even more patterns emerge in the detail of each field and paddock both from the ground and when viewed from high altitudes and even the edge of space. The patterns that we see look less mechanical and increasingly follow repeated living patterns that resemble roots, trunks, and branches. The well-intentioned assumption that simple answers are more likely to be correct undermines our ability to manage and see the beauty and resilience of complexity. The simplification of the complex stunts our ability to find, and appreciate, the patterns.

The complexity of managing systems that support regenerative outcomes on the land is part of the reason farmers, ranchers, and land stewards such as Amanda Cather and Emily Cornell have put so much effort into collaboration. It's what motivates me, as well. We all must rely on others who before us have fixed equipment, grazed animals in the woods, harvested oilseeds, and dried grain. Even with that help there are lifetimes of innovation ahead. We have each found that the land we steward will never reach its full potential for regeneration or resilience without the help of others, regardless of our individual efforts. And therein lies the paradox of mutually dependent independence.

In this chapter, we will explore scales of resolution, from the patterns observable at ground level to a more regional picture, where patterns of watersheds could be observable from a balloon all the way to the global satellite's view of the biosphere. In the process of moving from theory to practice, we might now visualize how individual efforts curated by people all around the world begin to intentionally connect and become observable even at the microbial level. It is a dizzying, empowering, and wondrous image of our world to comprehend. In my own lifetime, I have observed how my own farm has exhibited patterns like the growth rings of a tree. Seeing the fields I help steward become part of a larger pattern of a healthy watershed gives value to the sometimes hard and detailed work of managing soil health.

In looking at the larger landscape we can put our own "edges" in context as we see where our work ends and intersects and cross-pollinates with the stewardship of others, contributing to regional and even global outcomes. This agricultural mosaic of individual land stewardship efforts forms beautiful patterns and great works of art and restoration.

Beginning the Story from Ground Level

The smallest patterns of land stewardship start from an individual's observations, at eye level, and form the unique fingerprints they leave on the landscape. The ground-level patterns start microscopically but express themselves and change with our management of water, fire, plants, animals, harvests, and fertilization. It is at this level where land stewards' decisions are made that directly affect the billions of organisms in each handful of soil. It's where improvements to soil health are planned, practiced, measured, and adjusted. It is where businesses are built. It is the place where individuals implement practices informed by data, their own observations, and available

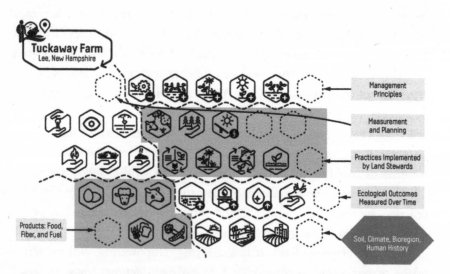

A Visual Language to Make Visible the Unique "Fingerprint" of People on All Land. An agricultural fingerprint, when represented in a visual language, enables a shareable, data-driven story about people and our influence on the land over time. Food, fiber, and fuel may represent just a small portion of the outputs of land management.

frameworks such as USDA soil health plans, holistic management principles, and permaculture designs.

The core principles of management are always at the biological level in relation to the land, be it a small plot or a thousand acres. It is this small piece of the picture that we think of as we look at the biological, physical, and chemical results of a soil health test and consider what micronutrients are required for a garlic, pear, or bean crop. We can observe in our own backyards how these ground-level patterns of regeneration build on the infinitely intricate details of nature. It is a process that we may all experience and which leaves not a footprint but a unique fingerprint on every product that comes from the land. At my own farm, I've had the privilege of adding my own fingerprint and using my hands to add relief and impressions to the land that can now be used as a unique identifier and tell a visual story of the origin of the products we make.

I have seen these patterns revealed over years as the permanent beds built over many growing seasons from mulched hay bales become more productive and require less water. I see the patterns revealed in the flush of growth in hay and pasture the season after the woodchips from forest management are spread on the fields as wood ash. I see the patterns revealed in the high bush blueberries and briars that provide shady grazing for the sheep flock in spring and fall and result in increased pollinators that reduce pest pressure. The patterns of diversity are built on the mosaics of nested activity, starting at the microbe level, and follow the same patterns of nutrient exchange from the field and forest to the garden. Grasses, legumes, brassicas, and diverse fungal populations of the sub-acre plots flow into the 100-acre managed woodland and silvopasture. Together, these represent the small patterns of personal experience and action at the level where we interact and experience the joys of observing and influencing changes in the small details of nature.

Ground-level patterns are found on farms in Malawi, on Emily Cornell's ranch in New Mexico, on Amanda Cather's Maryland farm, and everywhere else. The images they individually create are both unique and follow common patterns. Rather than thinking of a farm or ranch as creating carbon footprints, which are impressions left on the world from carbon extraction, we can think of these small patterns as carbon *footpaths* that have a destination, a purpose, and a story to tell. These carbon footpaths are visible in terraced hillsides, in lush hedgerows, in windbreaks, in the fire-managed understory of healthy forests, and in beaver meadows created where formerly there were ponds. They represent patterns of practices and can be expressed as both unique digital fingerprints and as stories shared through new knowledge utilities as part of our agricultural knowledge commons.

Science is not simply a tested hypothesis or even peer review, it is also an aggregated trust that enables each of us to build upon one another's knowledge. An isolated discovery is of no use if it cannot be communicated or understood. Additionally, a regenerative system will not work if we feel that our contributions are not being valued or that we are being manipulated or that value is being extricated without recognition, compensation, or consent. It is the movement from our own backyards to the watershed that our individual efforts take on new meaning.

The Watershed Level

As we navigate the patterns across watersheds with organizations such as the Million Acre Challenge, there is the constant social challenge to bridge across cultures, scales, geographies, and local production systems. At the medium level we elevate our perspective from the individual fingerprint of our own plots of land observable at ground level, to a higher altitude—say at the level of a hot-air balloon from 10,000 feet. From this aerial observatory we begin to see the patterns of communities and agroecology principles in action as the works of each individual are combined to form recognizable patterns of color and texture across millions of acres. It is here where the work of collaborative organizations such as the Million Acre Challenge in the Chesapeake Bay, the Quivira Coalition in New Mexico, the participatory research in Malawi, and knowledge sharing cooperatives in India begin to actually show up.

The medium scale is the work of collaboration and sharing of resources and exchanges of information. From this vantage point, Amanda Cather's farm and the farms across Maryland function as digital intermediaries within the OpenTEAM network. Hubs and their related regional agricultural networks serve as aggregators of trust and facilitate knowledge flow across a landscape. A hub is a point of connection from the local to the global. Hubs are the method by which OpenTEAM builds bridges between existing agricultural communities across diverse geographies, scales, languages, and cultures. We follow the threads and connections of an individual's steward-ship through the boundaries, both upstream and downstream, where ideas, information, and inspiration flow. For example we could make decisions about cover crops, agroforestry, silvopasture, reduced tillage, rotational graz-ing, or the many other innovations, both new and those of our ancestors, that are now being cross-pollinated, documented, adapted, and shared across the watershed.

Medium scale work is especially visible when we look across estuaries and watersheds. For example, let's further explore the group of farms emerging in Maryland from the Chesapeake Bay estuary, organized as the Million Acre Challenge. Together they are illustrating a simple choice to share knowl-edge. Each operation manages land individually, but they also have chosen to be part of the challenge, which designates each working farm as a Soil Health Hub. Together, they connect with other organizations regionally and nationally. Participation at this level does not compromise the sovereignty or autonomy of the individual enterprise, yet each farm has agreed to col-laborate to observe the biological flows of biodiversity, water, phosphorus, nitrogen, and carbon from their land while producing food, fiber, and fuel for nearby cities such as Baltimore and Washington, DC.

Chesapeake Bay is also part of a massive natural circulatory system with branches being fed by 100,000 rivers, streams, creeks, and tributaries stretching up to drainages in backyards, fields, and pastures. This same pat-tern applies universally, as to my local watershed in New Hampshire. In my backyard our patterns are organized around the Great Bay. Farther north in Portland, Maine, the Wolfe's Neck Center organizes its community around Casco Bay while collaborating with conservation districts south to the Con-necticut River watershed into Long Island Sound. These patterns persist from the Everglades to the Mississippi River Delta to the Nile, Amazon, and Yangtze Rivers.

As we move from our small-scale decisions of detail and autonomy to medium-scale decisions, the social and technical underpinnings of how decisions get made and who makes them shift. We are now not only man-aging the food, fiber, and fuel that leave our farm gate, as well as the soils, plants, and animals that produce them, but we are also being rewarded for stewarding public goods through the rivers that run through our lands.

The technical and social structures will adjust to support a different level of decision-making that starts to give voice and allocate new resources to nature.

Practically, we now have prototypes of the tools we need, employing knowledge utilities and voluntary networks such as the Million Acre Challenge to begin to create watershed-level, data-driven stories that bridge individuals, food systems, and a longer story of human history on the land. We can now create stories about our land that are both beautiful and true. Images and data can give body and voice to these stories as communities use digital tools to document and share change over time through watersheds, conservation districts, and agricultural organizations. It has been said that numbers move markets, but stories move people. By creating data-driven stories, we move both and can help make the invisible valuable. These stories show the role of nature at the scale of salt marshes, swamps, mountain water sources, rivers, and estuaries. There are new efforts being created by organizations such as the Regen Foundation, which seeks to create and endow tools that reestablish informal cultural ties with nature as a living entity involved in decision-making. By creating a public record, nature is given voice over time through the creation of knowledge utilities such as Public Land Library's catalog of soils and biodiversity.

On my own farm, the story being drafted and redrafted is about the role of the land being managed for water quality, particularly the Oyster River, which borders our farm and provides much of the drinking water for surrounding communities. Managing for water quality uncovers unexpected stories. For example, there is the story of turtles and trout, which thrive when wood snags slow the flow of the river and create deeper, shady, cool pools for protection and rest. It is a story of the fields and wooded edges, deep roots, and sunken logs that soak up nitrogen before the river reaches the Great Bay estuary's eel grass and oyster beds. It is a story of habitat being created for the New England cottontail and countless bees, wasps, and birds in the brushy margin at flowering field edges. It is the planting of pollinator meadows for migratory butterflies and beneficial insects, and the building of bat and bluebird houses. It is a story of deep-rooted clovers and radishes that break up compacted soil, bring nutrients from the atmosphere down to deeper soil layers, and pump minerals from the soil into the plants that are eaten at the surface. Each of these stories is about observing, listening, learning, and giving voice to the patterns of nature out of our shared observatories and part of increasing our commonwealth of nature.

Global Patterns of Stewardship

If the ground-level patterns are observable by individuals, and medium-level watershed patterns are observable from a balloon, then large-scale global patterns of agriculture are observable from the edge of the atmosphere, from weather balloons or low-Earth-orbit satellites. In the large-scale patterns of rivers and watersheds, geology and soils are put in context on their boundaries, and we can see the patterns of water, precipitation, migratory animals, and the effects of climate change emerge. Our global commonwealth of knowledge is where our scientific knowledge resides, where language and history are resolved as patterns of humanity on Earth. It is where nations and organizations at the continental level come together to share the story of our collective stewardship of our planet. It is captured in our Earth-orbiting satellites and in deep space observatories. It is expressed in data-driven stories that aggregate our collective triumphs and shortcomings by indicators, such as the United Nations' Sustainable Development Goals, and in stories we will be able to tell from a Public Land Library.

It was the launch of the early cameras into space, looking back at Earth, that gave us the first view of the global patterns of our land stewardship for the first time. When the *Voyager* spacecraft flew deep into the solar system, to barren Mars and boiling Venus, we gained a different appreciation for the patterns created from the unique distribution of water, oxygen, carbon, and nitrogen that make Earth a livable planet. The large patterns are where climate and biodiversity-accounting commitments are resolved as being effective or unmet, but are not where the work is done. The large patterns of land stewardship are also created by the work of the individual ground-level patterns and the coordination between neighbors. Watersheds cannot be acted upon directly or even seen directly by individuals. They must be assembled, plot by plot, into recognizable patterns of branches (in a process not unlike the individual pixels sent from the *Voyager* spacecraft looking back at Earth) and combined to render an image of our place in the solar system. Seeing and influencing the larger patterns is a process requiring the ongoing and long-term cooperation of millions of people.

One of our great challenges is that we must grapple with all three levels of scale simultaneously. To move toward a Great Regeneration, we must live and work in all three and build bridges across the boundaries. The global patterns of agriculture change much more slowly than the smaller ground-level fingerprints of individual plots or watersheds. The global patterns initially seem out of our control, even as we each contribute individually to the larger patterns that emerge. By democratizing access to observation tools such as Landsat imagery, radar maps, and maps of ocean currents, we are seeing just a hint of our commonwealth. When we can place our watersheds and landscapes in the larger patterns of change that are driving droughts and floods over the horizon, we can then initiate changes far more quickly in our own watersheds, fields, and gardens. While the Great Regeneration demands greater sovereignty, equity, and control at the individual level, it also demands greater sharing—not of scarce resources but of abundant ones.

We see the tension and transition of scale as we move from observing the autonomy of ground-level, individual places such as Amanda Cather's farm to the medium-scale collaboration of the Million Acre Challenge working across the Chesapeake Bay watershed to the even broader global context. The same watershed then contains parts of the largest patterns of migratory fish and birds, the warming gulfstream, a changing jet stream, rising and acidifying seas, and the collective effects of individuals on global climate regulation and adaptation.

Working at the global pattern level is a process of aggregating and maintaining transparency and trust, while not compromising individual liberty or sovereignty. It is a process of understanding the nuances of privacy, liberty, sovereignty, and autonomy at the individual management level, while tracking and creating tools to recognize value that transcends a watershed or an individual land steward's enterprise. It is the process of providing and endowing global knowledge utilities so that the smaller pieces may exist in context. The exchanges, standards, and services implemented at the individual and watershed levels should work in concert to provide the prerequisite for trusted data, knowledge, information, and inspiration for all. In turn, this will enable us to manage local markets, while also creating credible claims for regional-, national- and international-level climate markets, agreements, report cards, and benchmarks such as the UN Sustainable Development Goals, which we are increasingly using to track progress. It is at this global level particularly that governments and internet governance organizations—and their infrastructure—are important. It is at this global level that fiber-optic routes, server farms, and models for distributed hosting come into play. Like a sequoia, growth still happens at the cellular level, but each cell is coded within the context of growing a whole tree.

How might we develop a system that enforces sovereignty at the individual level, while also contributing to a shared knowledge base that is trusted, findable, accessible, interoperable, and reusable (FAIR)? This must remain a unifying question. We must remember this: If we are not in control of the tools we use, we are being controlled by them. If we can imagine an architecture to achieve sovereignty, trust, and the foundation of a global knowledge commons as the trunk of a mighty tree of humanity, then it seems like an obligation to create it.

As we traverse the scales of biology, social needs, and technical requirements, we see patterns build across the landscape, creating public goods that are transparent and private value that is stewarded within our boundaries. We see the individual sovereignty and individual needs reflected in the attention to detail in individual land stewardship. We see technical concepts working across these scales now being developed by OpenTEAM partners, who are manifesting knowledge utilities like the Public Land Library, as well as promoting autonomy, sovereignty, and control. However, ideas and information have value mostly when shared. Each of us has only part of the story to tell.

CHAPTER 9

Harvesting the Fruits of Our Labor

The state is a tree, agriculture its roots, population its trunk, arts and commerce its leaves. From the roots come the vivifying sap drawn up by multitudinous fibres from the soil. The leaves, the most brilliant part of the tree, are the least enduring. A storm may destroy them. But the sap will soon renew them if the roots maintain their vigour. If, however, some unfriendly insect attack the roots, then in vain do we wait for the sun and the dew to reanimate the withered trunk. To the roots must the remedy go, to let them expand and recover. If not, the tree will perish.

—MARQUIS DE MIRABEAU,
L'Ami Des Hommes, Avignon (1756)

As I came over a rise in the highway, night gave way and the sun burst through a gap in the ridgeline of Wyoming's Teton Range. Golden light bathed a jagged moonscape that was utterly at odds with the lush, soft hills of my family's New England farm. It was 2011, and I was four years into my dissertation at the University of New Hampshire while also farming with my wife full time. I was on my way to Jackson Hole to attend the wedding of my closest childhood friend, what should have been a four-hour trip. I had instead been traveling for twenty-six hours straight without sleep. I found myself driving a rental car overnight from Salt Lake City the last twelve hours of the journey as a last-ditch attempt to make it to the ceremony on time. Perhaps due to the combination of exhaustion, the grand wonder of the sunrise over the rugged landscape, the anticipation of seeing long-missed friends, and the

160 | The Great Regeneration

focus required to remain safely on the road, my mind entered a zone of clarity and expansiveness that I had never before experienced. I felt my perspective shift outside the car into the surrounding arid land, where I saw plants growing out of bare rock along the edges of the road. Moved by their struggle for survival, my consciousness and my recent studies merged with the environment as seasonal time and geological time blended together. I could visualize the efforts of plants and scrub brush to get a purchase on the rocks and soil, and the roots exploring cracks and pores of the rocky soil for limited moisture and minerals. Vastly different from forested and lush New Hampshire, this was a geologically young landscape where the weathering forces of wind and water fought the relentless uplift of mountains, while plants and animals managed year after year to survive in the arid conditions. It was not the first time I had been in this environment, but it was the first time I had the perspective to experience it in a new way. These ideas were no longer academic but flowed through me for a moment like breathing air into my lungs.

In that moment, the pieces of a puzzle I had been working on for a decade came crashing together all at once. I could more clearly visualize the cycles of carbon, water, and nitrogen in the atmosphere being captured and transformed into living plants across the landscape, and those plants adding oxygen to the atmosphere and feeding microorganisms below the soil. I saw the sun heating the surface of the land, creating the wind that evaporates water in the soil. Time both accelerated and slowed. I saw photosynthesis, the force behind nature's extraordinary capacity to regenerate itself and grow, counteracting the harshness of the landscape. I saw the geological forces both heaving the mountain ridge upward and also wearing it down. I saw my personal relationships and the upcoming marriage of my friends as an extension of the life-giving power we share with plants, part of the epic and eternal struggle to build complex and diverse webs that promote life even as the natural forces work to break them down and return the landscape to dust and minerals once more.

In my sleep-deprived yet enhanced mental state, the complex web of relationships forming a resilient ecosystem emerged as significant, present, and real—the product of millions of years of evolutionary forces, a wondrous process that had made plants designed to find a foothold wherever the sun shines. I could visualize the genetic code of soil microbes that had evolved to dissolve minerals and exchange them with plants to grow and transform the landscape. In that emotionally bare moment, I saw our human role as a purposeful search for understanding and taking responsible action on behalf of those natural processes. The wedding itself represented much more than the linking of two people but also the conjoining of relationships, networks, ideas, information, and inspiration.

In those first dramatic moments of feeling the Earth's rotation that revealed the sun over the horizon, agriculture and science took on a deeper meaning as a human endeavor to create regenerating, life-supporting order while enhancing the struggle of life against the chaos and entropy that eternally work to tear it all down. I saw agriculture as an incredibly intimate expression of our complex relationships with one another through food and ceremony as well as an expression of our practical curiosity to comprehend the world that we live in. With new clarity, I saw a responsibility toward understanding soil health, climate, and food systems as a unifying public science—a world being unlocked by telescopes, microscopes, satellites, digital cameras, and observational tools created out of our deepest shared curiosities. Decades ago, the astronomer Carl Sagan placed Earth in the context of a vast universe, using the symbol of a pale blue dot as the *Voyager* spacecraft turned to look back on its journey to the edge of our solar system. Witnessing the sunrise that day, I realized there was a similar *Voyager*-like moment just unfolding as we collectively began unlocking the secrets of the microbial and fungal universe beneath our feet.

As my thoughts that morning transported me through the rocky landscape, an ancient idea of progress and hope for the future began to emerge centered on a shared human pursuit of scientific understanding, social improvement, and civic engagement. It was a thought that would mature into a knowledge utility, a system of constantly flowing and expanding knowledge about animals, plants, and soils, gathered together and then redistributed for all to share as part of our commonwealth. I was enamored with the grand thought that a knowledge utility might also connect us all in a common conversation, transcending any particular language, geography, or culture. With shared science, all eight billion people on the planet could create any combination of ways to produce regeneratively abundant food, clean water, and healthy land.

When I arrived at the wedding, with tears of emotion still wet on my cheeks, I held on to the experience even as I walked from my car and made it to the ceremony on time. I have never experienced anything like it since, but I continue to draw on that experience and attempt to rediscover, see, and share with as much clarity as I had for those few minutes. The strength of the vision has since pushed me to pursue not just what is immediately expedient or practical, but to pursue the radical transformation that now seems possible.

Translating Revelation into Action

From that moment when the sunlight struck the Wyoming roadside and illuminated the struggle of life, I have attempted in many different ways to translate that revelation into action. Every year that I work with collaborative organizations like OpenTEAM, I can find a little more of the common journey of others, with glimpses of shared revelations and clarity as our efforts join together. Since that 2011 road trip, I became fascinated by the increasingly human-readable and intricate atomic-level patterns of hexagonally linked carbon structures that repeated across scales on the landscape and were being documented, copied, and shared. Our ancestors observed these same patterns, the difference now being that these patterns are readable and sharable in ways that enable us to tell new, more complex, and more exciting stories about all that is and will be.

The Wyoming sunrise experience, which was triggered by my observation of the struggle of a tree's tenacity to live and grow from the rocks, also unlocked a larger story of repeated patterns that started to take on new meaning for me. Upon returning to my work on the farm, it was clear that a deeper truth was missing in the academic field I was studying. I saw too much meaningless trust in competition, trust misplaced in the narrow framing of economics that was blind to the fundamental truth that the real competition is in an epic struggle for life and ecosystems versus a lifeless planet—not so much among species. After my Wyoming sunlight revelation, I became fixated on interpreting the story of the struggling roadside tree as a symbol for a type of irreducible complexity, which began to reveal patterns of human history across time and cultures.

When Marquis de Mirabeau described agriculture as the root of a great tree that encompassed industry, commerce, and generational collaboration of people with nature, he was tapping into a powerful and universal metaphor that suddenly made sense to me in a whole new way. After the Wyoming revelation, I could see and feel the flows and function of Mirabeau's tree in a way I think he might have seen it, but I also felt that I might see even further, with the benefit of 200 more years of growth having been added to the commonwealth of knowledge. I could now see the flow of information from agricultural roots to industry and commerce in the branches and leaves.

The revelation, combined with Mirabeau's insights and imagery, made me so much more curious. I felt I was tapping into a deeper, longer, universal human history rooted in the metaphor of the tree. Across cultures the tree has represented how we view our families' ancestries, the theory of evolution, and even the living and evolving structures of modern software code repositories. Suddenly, I could see the pattern of the "great tree" everywhere, from films such as *Avatar*, *Princess Mononoke*, and *Castle in the Sky* to its near omnipresence across religions and ancient and contemporary cultures, with variations repeated in different forms from Hungarian, Turkic, Mongol, Norse, Germanic, Slavic, Finnish, Baltic, Chinese, and Hindu mythology. I even began to see tree patterns in the rapidly growing open-source software communities that began to mirror the structures of roots, trunks, and branches. I could see reflections of the patterns repeated and codified in large-scale collaborative tools like GitHub and GitLab being used to create, govern, and maintain the software that runs nearly all information technology.

I came to connect the emotional experience of witnessing a small plant struggling to live a lifetime on a rocky expanse to the larger pattern of life as a cooperative effort very like the world tree. New start-ups, innovation, exploration, and competition are all expressions of seasonal leaves, whereas the process of seasonal growth and dieback is like an exploratory flow of ideas. We attempt to describe this truth in crude forms with our initial and imperfect attempts at a regenerative economy that is tied to natural climate solutions and the language embedded into the software we are creating.

By thinking about ourselves as part of a living structure that can grow deep roots to support more resilience, I have found meaning and happiness in living with autonomy, while sharing and improving what is common. We now see a need for rebalance with the requirement of rural broadband to support our silicon trunk of the commonwealth of knowledge. Without equal access to knowledge services, rural areas are like an arm with a tied tourniquet; without blood circulation the affected limb will eventually wither and die, to the detriment of the whole body. I have found that new organizations can be built to mirror these patterns of balance and values. Working with the Open-TEAM network, I see that projects layered and strengthened as ecosystems across the landscape contribute to the patterns of Mirabeau's tree and the larger pattern of competition of abundant life with lifeless geologic forces and weathering. These first attempts at growing the OpenTEAM networks are not random but purposeful, like new leaves growing at the tips of a large lattice of branches. These patterns of growth reveal living networks bound together by the forces of life on Earth over the forces of lifeless entropy.

The effort is not about capitalism, socialism, or any other "ism." There is a recognizable pattern of natural competition and collaboration as productive forces in tension, like branches on the same tree competing with others for light. It became clear that the emerging and natural balance between competition and collaboration was a pattern in natural systems that grow together. Without a vision of a shared whole of competition and collaboration in tension, industry grows out of balance.

Finding the balance is a search not about ideology—not what we *believe* works—but about patterns that *actually* work. We do not need to have a "belief" in gravity to fall when we trip. Like so many things, the questions we ask depend on what we already know. The solution in creating context is first to identify universal shared questions. For me, the Wyoming sunrise, and the pattern of our shared knowledge commons described by Mirabeau's tree, unlocked a new way of thinking about science and a shared process of searching for truth and asking more common questions to help reveal new paths as clear as a river on a landscape.

As we think about building durable structures to support a Great Regeneration, it is not simply about restoring small pockets of degraded landscape but about universal access to practices and patterns that grow solid and resilient structures capable of supporting broad exploration and experimentation. The shift to greater democratized access to agricultural knowledge is already creating tools that empower land stewardship, as we have seen throughout this book. The ability to provide integrity for environmental claims, as well as visibility for natural systems, is indeed increasingly within our grasp. Part of trust is creating shared language and the context for competition. It is hard to overstate the importance and potential rewards as we create networks of trust in a world that has otherwise resigned itself to content of dubious quality. Competition is a powerful exploration tool when it is harnessed within the commonwealth of nature as creative branches from a shared trunk.

A Land-Based Language

As Christopher Alexander advocated in *A Pattern Language*, languages of both the natural and built worlds can be interpreted and shared by empowered users to form workable large-scale solutions that evolve like living organisms. Although we have talked about the commonwealth of agricultural knowledge in broad terms, we have not yet discussed the language we use to communicate across boundaries. Without a new language of patterns to share, our commonwealth will be limited. In a sense we are in the midst of creating and extending a new pattern language to the large-scale architecture of our landscape through agriculture.

We do indeed have a pattern language of organization that starts at the molecular level, which we can manage and observe. We are just learning to hear and interpret it, like a baby beginning to imitate sounds from their parents and to understand meaning from gestures and sounds. We now know that trees and other plants are connected through vast fungal root networks and are able to communicate through a chemical language. As we expand our pattern language capacity, we might unlock the chemical codes of these messages to translate them into a digital language and then into a human-readable format. I sometimes think that my Wyoming revelation gave me just a flash of what fluency might feel like. I see the foundation of this new pattern language emerging from the branches of open-source code as a visual systems language of universal icons and symbols being created and curated to describe the unique fingerprint of stewarded land. The icons have individual integrity and meaning but can also be brought together like complex structures that form the cells that make up organisms. We now have visual-systems building blocks to describe and tell a new story of our land and stewardship.

One way to think about the work of OpenTEAM is an effort to share this new language by creating a layer of silicon mycelia to help exchange messages across the network that, in turn, grows our commonwealth of agricultural knowledge. Within OpenTEAM, and beyond, many attempts to create new pattern languages with visual, data-driven story platforms are being grown and explored by organizations mentioned in this book such as Terran Collective's Hylo project, and digital storytelling apps such as Lookin.to. Others, such as The Noun Project, are hosting pieces of a new icon-based visual language from which we can assemble our individual and common agricultural fingerprints and share observations and evidence in both our real coffee shops and in digital coffee shops being developed by organizations such as Our Sci for community research and exploration. It is in this context that patterns of agroecology reemerge to provide governing design principles to evaluate the outcomes of our work and our shared goals. Within those goals, there are examples and key functions that must be supported to help steward the environment to produce a healthy, balanced, and resilient knowledge commons that grows to meet our needs and strengthens itself in the process.

What are the questions we can answer together? We have the elements of abundance—both knowledge and natural resources—but they must be brought together and made visible. It is people, together, who will make the invisible visible and valuable. The people—the land and water stewards—are the ultimate intermediaries of our shared knowledge commons and how it is translated into action, but they are not alone. Land stewards are just the roots. Who will see themselves in this important role in the next generation?

We have key tools and an abundance of natural resources and human ingenuity to unlock nature's millions of years of ingenuity to translate an agroecology framework into action. OpenTEAM hubs function across cooperative agrarian organizations that are contributing to growing our commonwealth of agricultural knowledge. Agricultural community organizations are hubs of a larger social and technical web that is emerging and rooted in agrarian history—including the Quivira Coalition in New Mexico, the Maine Farmland Trust and Maine Organic Farmers and Gardeners Association, Pasa Sustainable Agriculture in Pennsylvania, the Million Acre Challenge in Maryland, the millions of farmers in cooperatives in India, the hundreds of thousands in farmers unions in the United States, Farmers Mutual in Australia, and cooperatives across South America. The land stewards in these growing networks have the elements of abundance to transform their local ecosystems through access to a global commonwealth.

As we look at the abundant resources across the landscape, we are also seeing vast sharable stories that embed the recipes for regeneration. However, a seed in the hand is not the same as a seed planted in the soil, and all the knowledge in the world about agriculture does little good if it is not applied through planting, pruning, and harvesting the fruits of our labor. Our shared potential must be translated into action. As we explore the potential to shift into the Great Regeneration, it often feels as if we have been collectively building a new racing bicycle without a chain. We have been coasting with it, which is faster than walking, but we have yet to add the final few pieces and step on a pedal for the first time. Although clearly the Great Regeneration is missing a few final pieces, we are perhaps closer than we think to moving at new speeds. And yet until those pieces are added we will be unable to reach our individual or shared potential we have already developed together.

Enjoying the Fruits of Our Labor

What might living in abundance feel like? I like to think about abundance as being a feeling of timelessness and casual foraging. I think about when I am able to eat a fresh peach I picked from trees planted with my parents ten years earlier. One special year I was able to pick and eat fresh peaches each day from July through September and had plenty to share. That struck me as a wonderful feeling of abundance. How might we expand that experience further to more aspects of our lives?

The history of that life force and abundance is all around us, if we choose to see it. The abundance of life is reflected not only in the soil, but even in the coal, oil, and gas that represents the billions of years of abundance and the rapid recycling of life. In our lifetimes we may observe the lives and deaths of organisms with shorter lifespans and, meanwhile, we might plant trees that will far outlive us. We can see branches in young trees that may wither after they have grown to provide energy to the roots. However, each version, each branch, each leaf that is created represents evolutionary exploration that helps move and energize the whole. The fruit of a tree represents the culmination and concentration of everything.

It is easy to forget that the fruit of our labor, especially in industry and commerce, is time we spend with one another, time to pursue our happiness and health, to think and create for ourselves and our families, and to experience and enjoy what might be, even as we see the world as it is. The fruit we produce is the reflection of our values, exchanged at the pace of family and friends and conversations over food while in the shade and safety of our vines and fig trees. We have only begun this conversation, and I have provided only a few examples from my own exploration. There are many more explorations ahead and more questions to ask. This book is an invitation to ask those questions and explore together. Do our questions need pruning to the core? Yes. Does our commonwealth of knowledge need fertilizing and repairing of damage? Yes!

We can now see that the limiting factor in the Great Regeneration is not access to carbon, nitrogen, water, or phosphorus but rather sharing our knowledge commons to ask better questions and manage for abundance, resilience, and biodiversity. This is about the potential for growth and change and for framing examples well beyond those covered by this book. As community-led organizations such as Practical Farmers of Iowa and California Farm Demonstration Network, and even the venture-backed Farmers Business Network, begin to form large-scale enterprises based on sharing trusted knowledge, it becomes ever more important to frame where we are collaborating and competing and recognize when a network can build a greater commonwealth together. If we view agriculture as limited by knowledge rather than inputs, then our focus also shifts to the questions we can begin to answer together.

When I revisit the sunlight emerging over the Teton Range again, I see a landscape transformed by the life of the land from a new perspective. I have seen extraordinary individual contributions made, and I am beginning to see the influence on patterns as more people increasingly see the whole of the shared tree of life and our growing commonwealth of knowledge. Certainly, we can't grow a Great Regeneration alone, but without doubt I am convinced we have the capacity to do it together.

An Agriculturalist Hippocratic Oath of Care

As agriculturalists and citizens, we are not responsible for feeding the world but for building a system by which individuals have the means to create their own abundance, feed their community, and take part in improving the system for their descendants. The promise of agriculture is not to build dependence but to create more independence, more choices, more liberty; by becoming more learned in this system, we may all become happier, living in a world whose systems we collectively understand to the extent that our senses allow. We partake in this human experience by attempting to transform these systems from the invisible into the visible, by bringing what had been shrouded in darkness into the light.

If we see our actions in the context of progress and civil society as having the common purpose of building a civilization that lives on the profits of nature justly created and enhanced, then our rights and privileges associated with agriculture should also follow. It is not a battle that is won but rather a cumulative, never-ending struggle in which progress can be accelerated through generations of collected knowledge.

Therefore, it is in our collective interests to lay out the points that unite us:

1. All members of civil society who are going to take action to improve and effect natural productive systems (agriculture) should have access to the best knowledge available on Earth, and it is in our shared interest to remove barriers to sharing this knowledge.

2. All knowledge of natural systems should be accessible, inexpensive, and shared widely.

3. Knowledge should be shared so as to improve its accuracy with every generation. As tools are a reflection of available systems knowledge, we have a right to access the best tools available based on our collective knowledge.

4. Potential knowledge of our environment is infinite, and each lifecycle helps illuminate patterns. Since our knowledge of the ever-changing environment also changes, all farmers and members of civil society should have the right to modify their tools to better suit their changing environmental conditions.

5. We have the right to use our collective body of knowledge to improve the productivity of natural systems, and where our labor has improved that system, the natural returns are then justly allocated based on our best understanding of science and art.

Agriculture is a human activity; therefore, its application and practice are as much a product of social-ecological systems and values as they are a product of technical or biological limits. The cultural and public nature of agriculture, however, is often understated or misrepresented as just another business, rather than the regenerative business that forms the basis for all others. Leonardo da Vinci is credited with saying, "We know more about the movement of celestial bodies than about the soil underfoot." And that is still true. Agriculture is a fundamental human pursuit and justifies the same public involvement that space exploration excites. What if we were to treat each farm like a newly discovered planet, with the same public curiosity, interest, and observational, analytical, and communications tools to understand our own biosphere and our own back fields? As we understand our environmental health as something we depend on for our well-being, we may internalize it as an extension of the health of our physical body.

Agriculturalists have an obligation that transcends the health of the individual. The agriculturalist's actions operate on the health of a system that affects not just a single patient but the health of all life. The physician, the plumber, the banker, the bus driver, the duck, the moose, the ant, and the mushroom all cycle water, carbon, and nitrogen, and all cycle through the soil that the farmer manages. Consequently, a new set of easy-to-remember principles needs codifying; they should be easy to teach and easy to hold people accountable to, just like the principles taught to doctors.

It seems high time that we extend to the agriculturalist a version of the physician's oath. In its original form, the Hippocratic oath requires a new physician to swear to uphold specific ethical standards. Of historic and traditional value, the oath is still considered a rite of passage for practitioners of medicine. But why only physicians? Why not holders of public data, why not engineers, why not scientists, why not businesses, and why not agriculturalists? What about the health and well-being of our atmosphere, our soils, our water, and our data about how the system of our biosphere functions and how we can create health or destroy it?

Because we, as agriculturalists, are able to contemplate our ability to improve or degrade biological systems health, we have a public obligation to uphold the highest standards of care. I suggest that we undertake this work, with credit to Diderot's 1751 *Encyclopédie*, to demonstrate the general system to the people with whom we live, and to transmit it to the people who will come after us, so that the works of centuries past are not useless to the centuries that follow—that our descendants, by becoming more learned, may become more virtuous and happier, and that we do not die without having merited being part of the human race. I will start by taking the oath myself (adapted from Dr. Louis Lasagna's 1964 Hippocratic oath).

I swear to fulfill, to the best of my ability and judgment, this covenant:

- I will respect the hard-won scientific gains of those farmers in whose steps I walk and gladly share such knowledge as is mine with those who are to follow.
- I will apply, for the benefit and health of all soil, all measures that are required, and not sacrifice long-term fertility for short-term yields.
- I will remember that there is art to farming as well as science, and that planning, observation, analysis, and collaboration with other farmers may outweigh the plow or the fertilizer hopper.
- I will not be ashamed to say "I do not know," nor will I fail to call in my colleagues when the skills of another are needed for the recovery of our soils.

- I will respect the privacy of other farmers, for their problems are not disclosed to me that the world may know. Most especially must I tread with care in matters of life and death.
- Above all, I must respect the power of nature and the power I have in daring to manipulate it for humanity's gains.
- I will remember that I do not treat an individual nutrient deficiency or a weed but a living soil whose wellness may affect not just our own economic stability but provide the basis of life for all future generations of humans and serve a crucial function for all life on Earth. My responsibility includes these related problems if I am to care adequately for the soil.
- I will use diversity to reduce disease whenever I can because prevention through increased photosynthesis, planning, systems, and understanding is preferable to fighting the life of the system through oversimplification.
- I will remember that I remain a member of society, with special obligations to all my fellow human beings to improve the most basic regenerative systems that support us all—those in cities and towns, and those in the country, and those sound of mind and body as well as the infirm.
- If and when I find my body unable to carry out these duties, I agree to pass along my accumulated knowledge, land, and resources to those who are willing and eager to apply their minds and bodies to uphold this oath.
- If I do not violate this oath, may I enjoy life and be respected while I live and be remembered with affection thereafter. May I always act so as to preserve the finest traditions of my calling, and may I long experience the joy of helping and improving our common soil with those who seek my help.

Key Terms and Resources

ADAPTIVE MANAGEMENT

Adaptive management is a process of continual improvement by adjusting "action" based on high-frequency observations and data-driven analysis rather than by expert opinion, best practice recipe, or prescription. It requires a high level of system understanding and observation, analysis, and communications feedback.

AG DATA WALLET

An Ag Data Wallet is a platform where land stewards can easily share their data across technologies, tools, and organizations while retaining control and data sovereignty. This means land stewards can determine which entities (if any) can access their data, how that data may be used or shared, and how that data may be aggregated. Through the Ag Data Wallet, land stewards can also revoke rights from any individual entity or specific use of their data, download data from the platform, or remove it entirely, and assign proxies to act as fiduciaries on their behalf.

AGROECOLOGY

Agroecology is defined by the Organization for Economic Co-operation and Development as "the study of the relation of agricultural crops and environment." Agroecology is the study of the interactions among plants, animals, humans, and the environment within agricultural systems. It is a holistic approach that seeks to reconcile agriculture and local communities with natural processes for the common benefit of nature and livelihoods.

AGROECOSYSTEM MODEL

A model for the functioning of an agricultural system, with all inputs and outputs. An ecosystem may be as small as a set of microbial interactions that take place on the surface of roots or as large as the globe. An agroecosystem may be at the level of the individual plant-soil-microorganism system, at the level of crops or herds of domesticated animals, at the level of farms or agricultural landscapes, or at the level of entire agricultural economies.

ANALYTIC TOOL

An analytic tool is any tool used to interpret and make use of observations and observational data. Examples include tables, charts, statistical tools (spreadsheets, for example), and models used to predict future system behavior based on past observations.

BACKBONE ORGANIZATION

A backbone organization is a separate organization dedicated to coordinating the various dimensions and collaborators involved in an initiative. A defining feature of the collective impact approach, backbone infrastructure is essential to ensuring the effort maintains momentum and facilitates impact.

CARE PRINCIPLES

The CARE Principles for Indigenous Data Governance were created to advance the legal principles underlying collective and individual data rights in the context of the United Nations Declaration on the Rights of Indigenous Peoples (UNDRIP). CARE is an acronym that stands for collective benefit, authority to control, responsibility, and ethics.

While CARE can be considered part of the open-data movement, it aims to build on other standards such as FAIR (findable, accessible, interoperable, reusable) by considering power differentials and historical contexts. The CARE Principles for Indigenous Data Governance are "people and purpose-oriented, reflecting the crucial role of data in advancing Indigenous innovation and self-determination."

COLLECTIVE IMPACT

Collective impact is the commitment of a group of actors from different sectors to a common agenda for solving a specific social problem, using a structured form of collaboration. Successful collective impact initiatives typically have five conditions that together produce true alignment and lead to powerful results: a common agenda, shared measurement systems, mutually reinforcing activities, continuous communication, and backbone organizations.

COLLABATHON

A compound word created by combining "collaboration" with "marathon," it is a sustained design and devlopment effort with short sprints toward long-term shared goals.

COMMONING

As a verb, commoning serves to emphasize an understanding of the commons as a process and a practice rather than as "a particular kind of thing" or static entity. There is no commons without the act of commoning—the practices we use to manage a resource and the values that guide them. The goal of commoning is to ensure collective benefit for humanity and nature.

COMMONS

The commons is the cultural and natural resources accessible to all members of a society, including natural materials such as air, water, and a habitable Earth. These resources are held in common even when owned privately or publicly. Commons can also be understood to refer to natural resources that groups of people (communities, user groups) manage for individual and collective benefit. Characteristically, this involves a variety of informal norms and values (social practice) employed for a governance mechanism. Commons can also be defined as a social practice of governing a resource not by state or market but by a community of users that self-governs the resource through institutions that it creates.

COMMONWEALTH

Originally a compound word, common-wealth, it comes from the old meaning of "wealth," which is "well-being," and literally means "common well-being." In the seventeenth century, the definition of "commonwealth" expanded from its original sense of "public welfare" to include forms of governance. According to World Bank economist Herman Daly, commonwealth is "wealth that no one has made, or the wealth that practically everyone has made. So it's either nature—nobody made it, we all inherited it—or knowledge—everybody contributed to making it, but everyone's contribution is small in relation to the total and depends on the contributions of others."

COMMUNITY OF PRACTICE

A community of practice (CoP) is a group of people who share a craft or a profession. It is through the process of sharing information and experiences with the group that members learn from one another and have an opportunity to develop both personally and professionally.

CREATIVE COMMONS LICENSE

A Creative Commons (CC) license is one of several public copyright licenses that enable the free distribution of an otherwise copyrighted work. A CC license is used when an author wants to give people the right to share, use, and build upon a work that they have created.

DATA SOVEREIGNTY

Data sovereignty is usually referenced in the context of an individual's ability to fully create and control their credentials, identity, and related information about themselves and their work. Data sovereignty is seen by Indigenous peoples and activists as a key piece to self-governance structures and an important pillar of food sovereignty and sovereignty as a whole. With the rise of cloud computing, many countries have passed various laws around control and storage of data, which all reflects measures of data sovereignty. More than 100 countries have some sort of data sovereignty laws in place.

DECISION SUPPORT SYSTEMS

The terms Decision Support Systems (DSS) and Decision Support Tools (DST) refer to a wide range of computer-based tools (simulation models or techniques and methods) developed to support decision analysis and participatory processes. A DSS consists of a database and different tools coupled with models, and it is provided with a dedicated interface in order to be directly and more easily accessible by nonspecialists (that is, decision-makers). DSS have specific simulation and prediction capabilities and are also used as vehicles of communication, training, and experimentation. Principally, DSS can facilitate dialogue and exchange of information, thus providing insights to nonexperts and supporting them in the exploration of management and policy options.

DIRECT MEASUREMENT

Direct measurement is when the characteristics of something can be explicitly measured rather than using indicators, models, or proxies. Examples of tools for direct measurement of the environment include wind vanes, anemometers, thermometers, and rain gauges.

DISTRIBUTED LEDGER

A distributed ledger is a consensus of replicated, shared, and synchronized digital data geographically spread across multiple sites, countries, or institutions. There is no central administrator or centralized data storage.

ECOSYSTEM SERVICE MARKETPLACE

Ecosystem services are the many and varied benefits that humans freely gain from the natural environment and from properly functioning ecosystems such as air and water quality, habitat, esthetics, and recreation. A marketplace quantifies and creates markets based on the change over time in those services.

ENVIRONMENTAL CLAIMS CLEARINGHOUSE

The purpose of an environmental claims clearinghouse (ECC) is to allow the broader market to continue to evolve, while bringing trust and stability to certain aspects of the day-to-day creation and exchange of these claims. An operational clearinghouse would enable the development of new and diverse claim asset classes across the world while providing a trusted methodology for claim identification and assurance of uniqueness. An ECC is similar to a government property clearinghouse or land registry. By providing a searchable database of land ownership or tenancy, the sale of land interests is simplified, allowing purchasers to be confident that the seller has not already sold to another buyer. It further allows the differentiation of rights, for example the right to pump water from underneath the land versus the right to build and live in a structure on the surface. An ECC also gives a claim purchaser confidence that they are the sole purchaser of a particular claim, while identifying other claims which might be relevant.

FAIR DATA PRINCIPLES
FAIR data follow the principles of findability, accessibility, interoperability, and reusability. The acronym and principles were defined in a March 2016 paper in the journal *Scientific Data* by a consortium of scientists and organizations. The FAIR principles emphasize machine-actionability (that is, the capacity of computational systems to find, access, interoperate, and reuse data with minimal or no human intervention) because humans increasingly rely on computational support to deal with data as a result of the increase in volume, complexity, and creation speed.

FARM HACK
Farm Hack is a community of collaborators interested in developing and sharing open-source tools for resilient agriculture. Individuals and organizations, nonprofits and businesses alike are invited to participate in the platform for community-based sharing and collaborative research. Through an open-source ethic, Farm Hack aims to retool farms for a sustainable future.

FARMOS
FarmOS is a free, open-source, web-based farm management and record-keeping system. It also provides a platform for the creation of custom modules and integration across decision tools, observation tools, and agricultural Internet of Things devices.

FOOD SOVEREIGNTY
Food sovereignty is a system in which the people who produce, distribute, and consume food also control the mechanisms and policies of food production and distribution.

FREE AND OPEN-SOURCE SOFTWARE
Free and open-source software (FOSS) is freely licensed to anyone to use, copy, study, and change in any way. The source code is openly shared, and people are encouraged to voluntarily improve the design of the software.

GATHERING FOR OPEN AGRICULTURE TECHNOLOGY
The Gathering for Open Agriculture Technology (GOAT) is a community of practice as well as a conference to identify what exists and what is missing

in the open agricultural technology landscape. Their mission: "The technologies that produce our food and the data about our food system should be public, and enable control by the farms and farmers that produce it. Together, we can collectively address the problems which prevent the creation of advanced, high quality open technology and its adoption."

GATHERING FOR OPEN SCIENCE HARDWARE

The Gathering for Open Science Hardware (GOSH) is a movement that seeks to reduce barriers between diverse creators and users of scientific tools to support the pursuit and growth of knowledge. GOSH principles are to be accessible, make science better, be ethical, change the culture of science, democratize science, have no high priests, empower people, have no black boxes, create impactful tools, allow multiple futures for science.

GNU GENERAL PUBLIC LICENSE

The GNU General Public License (GPL) is a computer software copyleft license. A copyleft license lets the user of the software use a program in many of the same ways as if it were public domain. They can use it, change it, and copy it. They can also sell or give away copies of the program with or without any changes they made to it. The license lets them do this as long as they agree to follow the terms of the license.

HACK VS. TOOL

A hack is an unusual assembly of available components to address a particular challenge. It is an individual effort and creates an isolated workable solution and is the basis for empowerment and innovation using global knowledge and local production. A tool is any workable hack that has been tested and replicated over time and by other parties. A hack becomes a tool through documentation and communication. It can be a physical object or a method or framework that can be documented (software, for example).

INTEROPERABILITY

Interoperability is a characteristic of a product or system whose interfaces are completely understood to work with other products or systems, at present or in the future, in implementation or access, and without any restrictions.

While the term was initially defined for information technology or systems engineering services to allow for information exchange, a broader definition takes into account social, political, and organizational factors that impact system-to-system performance.

INTERNET OF THINGS

The Internet of Things (IOT) is the network of physical devices and other items embedded within electronics, software, sensors, actuators, and connectivity that enables these things to exchange data, creating opportunities for more direct integration of the physical world into computer-based systems.

LANDPKS

The Land Potential Knowledge System (LandPKS) is a simple, free, and open-source software tool and mobile application developed by USDA Agricultural Research Service that supports land-use planning and management. It includes modules that allow non–soil scientists to (a) determine the sustainable potential of their land (LandInfo), (b) monitor the health of their land (LandCover and SoilHealth), and (c) record management activities (LandManagement). The land potential assessments will be updated based on new evidence regarding the success or failure of new management systems on different soils. The knowledge engine together with mobile phone applications and cloud computing technologies facilitate more rapid and complete integration of local and scientific knowledge into land management.

MODULAR DESIGN

Modular design is a design approach that subdivides a system into smaller parts called modules that can be independently created and then used in different systems. A modular system can be characterized by functional partitioning into discrete, scalable, reusable modules; rigorously using well-defined modular interfaces; and making use of industry standards for interfaces.

OBSERVATIONAL TOOL

An observation tool is used to record behavior over time and describe the conditions of change. Observation technology also enhances the basic human senses to enable greater perception into the environment (how things work). Examples include qualitative tools such as cameras, microscopes, telescopes, and thermal imaging, as well as quantitative tools with analog and digital sensors such as thermometers, barometers, and pH, soil, and air-moisture sensors.

ONTOLOGY

An ontology is a set of concepts and categories in a subject area or domain that shows their properties and the relations between them. Shared data ontologies, alongside shared vocabularies, are used to support the interoperability of software and tools.

OPEN DATA

Open data is data that can be freely used, reused, and redistributed by anyone—subject only, at most, to the requirement to attribute and share alike. The core of a "commons" of data (or code) is that one piece of "open" material contained therein can be freely intermixed with other "open" material. This interoperability dramatically enhances the ability to combine different datasets together and thereby to develop more and better products and services.

OPEN SOURCE

Open source is a publicly accessible software or hardware design that can be modified and shared by multiple users. This allows the source code to be inspected and enhanced by anyone. Because open source uses multiple collaborators, it allows for more control, increased security and stability, additional training opportunities, and the foundation of communities centered on software design.

OPEN-SOURCE APPROACH

The open-source approach encompasses a broad set of values and principles derived from open-source software development. This approach is centered on transparency, collaboration, inclusivity, and community. By including open access, collaboration among diverse perspectives, and rapid prototyping, the highly iterative open-source approach produces an optimized product generated by communities.

OPENTEAM

OpenTEAM (Open Technology Ecosystem for Agricultural Management) is a network of public and private partners who aim to collaboratively develop and maintain an ecosystem of tools and networks that use low-cost, open-source technologies that link producers, researchers, businesses, and the public. It uses a shared, open architecture that leverages existing capacity and enables exchange of agricultural information and inspiration across geographies, production systems, and cultural boundaries.

OUR SCI

Our Sci is a collaborative research platform that uses open-source tools to create customizable surveys, collect data, and report and share results. OpenTEAM specifically uses Our Sci's SurveyStack application to create customizable surveys that can be directly integrated into farmOS.

PARTICIPATORY ACTION RESEARCH

Participatory action research (PAR) is an approach to research in communities that emphasizes participation and action. It seeks to understand the world by trying to change it collaboratively and following reflection. PAR emphasizes collective inquiry and experimentation grounded in experience and social history.

PRECOMPETITIVE COLLABORATION

Precompetitive collaboration uses a collective approach to bring together a diverse group of stakeholders to create new technologies and solutions benefiting a shared industry. Diverse efforts can be streamlined into solving systematic problems using multiple industry perspectives.

PUBLIC LAND LIBRARY

A Public Land Library is a publicly curated knowledge commons comprising a curated set of environmental, geological, and human history records of land that is documented and accessible. Like a traditional public library, its documents and archives can be contributed to, and they are searchable like a registry of deeds, while also protecting sensitive locations or data that might threaten food sovereignty.

REGENERATIVE AGRICULTURE

In biology, regeneration is the process of renewal, restoration, and growth that makes cells, organisms, and ecosystems resilient to natural fluctuations or events that cause disturbance or damage. Regenerative agriculture is a system of farming principles and practices that increases biodiversity, enriches soils, improves watersheds, and enhances ecosystem services.

REPLICATION VS. DUPLICATION OF EFFORTS

Replication of efforts is the conscious process of repeating the efforts of others and communicating the results of those efforts. Duplication of efforts is the undesired state of isolated problem-solving without knowledge of prior or parallel work, or without conscious communication of the results of those efforts.

RIGHTS OF NATURE

Rights of nature, or Earth rights, is a legal and jurisprudential theory that describes inherent rights as associated with ecosystems and species, similar to the concept of fundamental human rights. The rights of nature concept challenges twentieth-century laws as generally grounded in a flawed frame of nature as a "resource" to be owned, used, and degraded. Proponents argue that laws grounded in rights of nature direct humanity to act appropriately and in a way consistent with modern, system-based science, which demonstrates that humans and the natural world are fundamentally interconnected. They take the form of constitutional provisions, treaty agreements, statutes, local ordinances, and court decisions. As of 2021, rights-of-nature laws exist at the local to national levels in seventeen countries, including dozens of cities and counties throughout the United States.

SHARED LIBRARY

In computer science, a shared library is a collection of nonvolatile resources used by computer programs, often for software development. These can include shared data, documentation, and prewritten code that can be common across many projects and therefore does not need to be recreated.

SMART CONTRACT

Smart contract might better be referred to as an auto-executable contract. It is a computer protocol intended to digitally facilitate, verify, or enforce the negotiation or performance of a contract. Smart contracts allow the performance of credible transactions without third parties. These transactions are trackable and irreversible.

SOFTWARE DEVELOPMENT KIT

A software development kit (SDK) is a collection of modules that can be adaptively deployed to more rapidly develop functional applications without having to build them from scratch. For example, Apple has an SDK for developing for iOS, and Android has a Java development SDK. An SDK can take the form of a simple implementation of one or more application programming interfaces (APIs) in the form of on-device libraries to interface to a particular programming language, or it may be as complex as hardware-specific tools that can communicate with a particular embedded system. Common tools include debugging facilities and other utilities, often presented in an integrated development environment (IDE). SDKs may also include sample code and technical notes or other supporting documentation such as tutorials to help clarify points made by the primary reference material.

SOFTWARE PLATFORM

A software platform is the environment in which a piece of software is executed. It may be the hardware or the operating system (OS), even a web browser and associated application programming interfaces (APIs), or other underlying software, as long as the program code is executed with it. Computing platforms have different abstraction levels, including a computer architecture, an OS, or runtime libraries. A computing platform is the stage on which computer programs can run.

SOIL HEALTH INDICATORS

Soil health indicators are measurable properties of soil or plants that provide clues about how well the soil can function. Indicators can be physical, chemical, or biological properties, processes, or characteristics of soils. They can also be morphological or visual features of plants.

SOILSTACK

A growing academic research initiative and protocol for rapidly assessing soil carbon content across landscapes. SoilStack creates an accessible measurement system that empowers individuals to generate reliable soils carbon data for the purpose of ecological understanding, decision-making, and markets. By providing an inexpensive avenue for measuring soil carbon contents, SoilStack allows for researchers to collect hundreds of measurements across landscapes.

SURVEYSTACK

SurveyStack, developed by Our Sci, is an open-source, customizable data collection tool to create surveys that can be directly integrated into farmOS. Users collect data using an Android app and Bluetooth or USB connection to attach measurement data from instruments.

UTILITY

A utility is used to support the general functions of software infrastructure in contrast to application software, which is aimed at directly performing tasks that benefit ordinary users. A utility maintains the infrastructure for a public service and also provides a service using that infrastructure. In our software context, utilities can also refer to a common set of microservices and services provided by the community, including but not limited to weather data, soil data, geographic information system (GIS) layers.

Acknowledgments

Special thanks to Courtney White for the encouragement, friendship, long explorations, and patience over many years without which this book would not exist; and to Ben Trollinger for the guidance and editing. I also owe a great debt to more people than I can name who shaped conversations and collaboration over the last ten years, including but not limited to Bianca Moebius-Clune, Jo Guldi, Don Blair, Tim Cooke, Tim Tenson, Tom Kelly, Greg Landau, Rikin Gandhi, Julien Reynier, Jan-Hendrik Cropp, Bob Eckert, Rich Smith, Abe Collins, Peter Donovan, Brandon Smith, Laura Gilmer, Shefali Mehta, Laura Morton, Klaas Martens, Sallie Calhoun, Dennis Meadows, Bill Coperthwaite, David Bollier, Greg Watson, Lakisha Odum, Britt Lundgren, Jeff Herrick, Sieg Snapp, Todd Barker, Dave Herring, Greg Austic, Mike Stenta, Jeff Warren, RJ Steinert, Severine Flemming, Marcin Jakubowski, Chelsea Cary, Jessica Ciartis, Dan Kane, Phil Taylor, Bill Buckner, Mike Comp, Lance Gunderson, Zach Wolfe, Richard Teague, Dan Kittredge, and Aaron Ault. I also owe deep gratitude to the welcoming open-source communities of GOSH, GOAT, and OpenTEAM, and Wolfe's Neck Center for Agriculture and the Environment staff for helping to shape a hopeful vision while passionately translating theory into practice every day. Truly, my deepest gratitude.

—DORN COX

My work on this book would not have been possible without the financial support of the following: The Bradshaw-Knight Foundation, Bybee Foundation, Globetrotter Fund, Grasslands Foundation, the O'Toole Family Foundation, The Regenerative Agriculture Foundation, and Ron Johnson and Brandie Hardman Johnson from Boulder Mountain Guest Ranch with Singing Earth Foundation. I deeply appreciate the support of every individual and organization.

—COURTNEY WHITE

Notes

INTRODUCTION

1. Tom Atlee, "Crisis Fatigue and the Co-Creation of Positive Possibilities," *The Co-Intelligence Institute*, https://www.co-intelligence.org/crisis _fatigue.html.
2. Elizabeth Kolbert, *The Sixth Extinction: An Unnatural History* (New York: Henry Holt, 2014).
3. Gwyn Jones and Chris Garforth, "The History, Development, and Future of Agricultural Extension," *UN Food & Agriculture Organization*, https://www.fao.org/3/W5830E/w5830e03.htm.
4. Gabe Brown, *Dirt to Soil: One Family's Journey into Regenerative Agriculture* (White River Junction, VT: Chelsea Green, 2018).
5. IPCC, *Special Report: Climate Change and Land*, 2019, https://www.ipcc .ch/srccl.
6. David Montgomery, *Dirt: The Erosion of Civilizations* (Berkeley, CA: University of California Press, 2012).
7. IPCC, *Special Report* 2019.

CHAPTER 1. THE GOOD ANTHROPOCENE

1. Jan Zalasiewicz, et al., "When Did the Anthropocene Begin? A Mid-Twentieth Century Boundary Level Is Stratigraphically Optimal," *Quaternary International* 383 (October 5, 2015): 196–203, https://doi .org/10.1016/j.quaint.2014.11.045.
2. Alex Aves, et al., "First Evidence of Microplastics in Antarctic Snow," *The Cryosphere* 16, no. 6 (June 7, 2022): 2127–45, https://doi.org/10.5194/ tc-16-2127-2022.
3. IPCC, *Special Report*, 2019.
4. "The Human Epoch," Editorial, *Nature* 473, 254 (May 2011): https://doi .org/10.1038/473254a.

5. Paul Crutzen and Christian Schwägerl, "Living in the Anthropocene: Toward a New Global Ethos," *Yale Environmental 360*, January 24, 2011, https://e360.yale.edu/features/living_in_the_anthropocene_toward _a_new_global_ethos.

6. Tim Mullaney, "Jobs Fight: Haves vs. the Have-Nots," *USA Today*, September 16, 2012.

7. Kae Tempest, "Parables," 2014, https://lyrics.lol/artist/196091-kate -tempest/lyrics/4639614-parables.

8. Clive Hamilton, "The Theodicy of the 'Good Anthropocene,'" Duke University Press, *Environmental Humanities* 7, no. 1 (May 1, 2016): 233–38, https://doi.org/10.1215/22011919-3616434.

9. Nibedita Mohanta, "How Many Satellites Are Orbiting the Earth in 2021?," *GeoSpatial World*, May 28, 2021, https://www.geospatialworld .net/blogs/how-many-satellites-are-orbiting-the-earth-in-2021.

10. "Landsat Then and Now," NASA, Landsat Science, https://landsat .gsfc.nasa.gov/about.

11. "Landsat," NASA, Landsat Science.

12. "Planetary Dashboard Shows 'Great Acceleration' in Human Activity Since 1950," Press release, IGBP, January 15, 2015, http://www .igbp.net/news/pressreleases/pressreleases/planetarydashboard showsgreataccelerationinhumanactivitysince1950.5.950c2fa1495db 7081eb42.html.

13. Montgomery, *Dirt*.

14. "UN Report: Nature's Dangerous Decline 'Unprecedented'; Species Extinction Rates 'Accelerating,'" UN Sustainable Goals, May 6, 2019, https://www.un.org/sustainabledevelopment/blog/2019/05/nature -decline-unprecedented-report/.

15. "UN Report: Nature's Dangerous Decline," UN Sustainable Goals.

CHAPTER 2. OUR COMMONWEALTH OF KNOWLEDGE

1. David Bollier, "The Commons, Short and Sweet," blog, July 15, 2011, http://www.bollier.org/commons-short-and-sweet.

CHAPTER 3. PUBLIC SCIENCE AND SOIL HEALTH

1. Jo Guldi, *The Long Land War: The Global Struggle for Occupancy Rights* (New Haven, CT: Yale University Press, 2022).

2. L. Winner, "Do Artifacts Have Politics?" from *The Whale and the Reactor: A Search for Limits in an Age of High Technology* (Chicago: University of Chicago Press, 1986): 19–39, https://www.cyut.edu.tw/~ckhung/b /sts/winner.html.

CHAPTER 5. THE ART AND SCIENCE OF COLLABORATION

1. David Graeber and David Wengrow, *The Dawn of Everything: A New History of Humanity* (New York: Farrar, Straus and Giroux, 2021).
2. Food and Agriculture Organization of the United Nations, "2nd International Symposium on Agroecology: Scaling up agroecology to achieve the Sustainable Development Goals (SDGs), 3–5 April 2018, Rome," https://www.fao.org/3/CA0346EN/ca0346en.pdf.

CHAPTER 6. THE ELEMENTS OF ABUNDANCE

1. Christopher J. Rhodes, "Solar Energy: Principles and Possibilities." *Science Progress* 93, no. 1 (2010): 37–112, http://www.jstor.org /stable/43424235.
2. Nic Carter, "How Much Energy Does Bitcoin Actually Consume?" *Harvard Business Review*, May 5, 2021, https://hbr.org/2021/05/how -much-energy-does-bitcoin-actually-consume.
3. USDA Natural Resources Conservation Service, "Soil Health," https:// www.nrcs.usda.gov/conservation-basics/natural-resource-concerns /soils/soil-health.
4. *The Hugh Bennett Lectures*. Raleigh, North Carolina: The Agricultural Foundation, Inc., North Carolina State College, June 1959.

CHAPTER 7. THE TECHNOLOGY OF TRUST

1. Guldi, *The Long Land War*.
2. Dana L. Thompson, et al., "The Importance of Design in Lithium Ion Battery Recycling—A Critical Review," *Green Chemistry*, vol. 2020, no. 22, pp. 7585–603. https://doi.org/10.1039/d0gc02745f.

Index

About the Authors

DORN COX is the research director for the Wolfe's Neck Center for Agriculture and the Environment in Freeport, Maine, and farms with his family on 250 acres in Lee, New Hampshire. He is a founder of the farmOS software platform and Farm Hack and is active in the soil health movement. In 2018 he received the inaugural Hugh Hammond Bennett Award for Conservation Excellence given by the National Conservation Planning Partnership. In 2019 he won a Ground-Breaker Prize from FoodShot Global for his leadership in developing the Open Technology Ecosystem for Agricultural Management (OpenTEAM). He speaks regularly about participatory science, open agricultural-knowledge exchange, and regenerative agriculture. He has a BS from Cornell University and a PhD from the University of New Hampshire in natural resources and Earth systems science.

Tuckaway Farm

COURTNEY WHITE is a former archaeologist and Sierra Club activist who dropped out of the "conflict industry" to cofound the Quivira Coalition, a nonprofit conservation organization dedicated to building a radical center among ranchers, conservationists, and public land managers around practices that improve resilience in Western working landscapes. In 2005, Wendell Berry included Courtney's essay "The Working Wilderness" in his collection titled *The Way of Ignorance*. He is the author of *Revolution on the Range*; *Grass, Soil, Hope*; *The Age of Consequences*; and *Two Percent Solutions for the Planet*; and coauthor of *Fibershed* with Rebecca Burgess. He is also the author of *The Sun*, a mystery novel set on a working cattle ranch in northern New Mexico. He lives in Santa Fe.

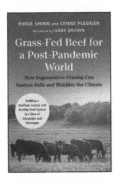

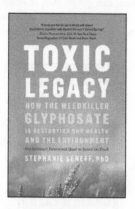
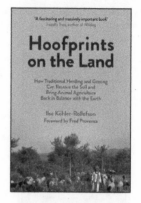

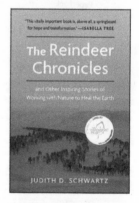